C0-BWI-996

VANISHING WILDERNESS

GIRAFFE

VANISHING WILDERNESS

by

F. R. La Monte and M. H. Welch
of The American Museum of Natural History

Illustrations
by
CAPTAIN VLADIMIR PERFILIEFF

Foreword
by
DR. JAMES L. CLARK
VICE-DIRECTOR IN CHARGE OF PREPARATION AND EXHIBITION,
AMERICAN MUSEUM OF NATURAL HISTORY

LIVERIGHT PUBLISHING CORPORATION
NEW YORK

* * * *

BLACK AND GOLD EDITION

1949

MANUFACTURED IN THE U. S. A.

FOREWORD

BEFORE all too short a time, a wild animal, roaming freely in our open wilderness, may be a thing of the past. When I first visited Africa twenty-five years ago, tremendous herds of wild game roamed the veldt; but now, when I return, every five years or so, the "animal map" has vastly changed. Where I saw elephants, stands a railroad station. Where I saw countless numbers of zebra and hartebeest peacefully grazing, airplanes now hover and settle to leveled and well-kept landing fields.

The transmutation of the wilderness into cultivated farm lands has robbed these creatures of their very home. With man's every new invention comes his certain and easier invasion of their homes and sanctuaries. Animals like the eland, the zebras and some of the buffaloes, which may yield to domestication, may, by this alone, withstand the challenge of extermination. But others, even though they may be harmless, are in man's way in his altruistic development of this old world, and are being crowded back into corners unsuitable to their habits and certain to their doom.

Throughout the world man is pushing forward. Animals are moving back.

The animals of to-day, whose lives have been, after all, so much a part of our ancestors' and ours, too, if we but realize it, may be the fossils of another day—a day which is coming all too soon, coming faster than we realize—when habits and haunts of our wild creatures are but memories recorded in books like this, cherished and preserved, for they were written by those who *knew,* back in those wonderful days when wild animals once lived upon our earth.

JAMES L. CLARK

Vice-Director (In Charge of Preparation and Exhibition)

The American Museum of Natural History, New York.

TABLE OF CONTENTS

LIST OF ILLUSTRATIONS

VANISHING WILDERNESS

1

THE AMERICAN BISON

THE early sea-rovers who first discovered our country were bitterly disappointed with what they found. They had risked their lives, tossing on the stormy Atlantic in their rocking ships for months. During the long, rough voyage, they bolstered up their courage with talk of the treasure that awaited them on the shores of the New World. They dreamed of great cities, of glittering palaces, of unmeasured riches—all theirs for the taking. They reached safe harbor, and disembarked, thankful for the feel of firm ground beneath their feet. They strode inland with hearts bounding and hopes high.

Alas for their dreams! They found no cities, no palaces, no gold. They found only bleak coasts and rugged forests. The New World was a wilderness.

Sullen and angry, the sailors refused to remain. They cursed their captains for their treachery in leading them to so uncouth and empty a land. They demanded a safe return home. Under the rebellious mutterings and threats of their men, the captains lost heart. Some of them tried to persuade their crews to stay and build forts and settlements, but the hardships of the winter, the dread of Indians lurking in the woods and the dismay of loneliness defeated them. In the end, they yielded to discouragement. Once more the ships were manned, and the sea-rovers gave themselves to the waves again, homeward bound.

So, over a period of several hundred years, crew after crew set foot on America's eastern shores—and sailed away.

During their brief encampments on land, they reconnoitered the forests along the coast. They made the acquaintance of the wild life of the woods. They followed the track of the deer, and of the bear. They caught the silken squirrel and the sharp-quilled porcupine, the woodchuck, the badger, the beaver, wild turkeys and the multitude of other creatures that provided them with food for their pots and furs for their bodies.

But of America's largest and most numerous game, these first-comers saw nothing. Beyond the rim of the forest and mountain barriers of the East, spread the vast prairies—the home of the American bison. The plains were black with the huge herds. All the way from the Alleghany Mountains to the wall of the Rockies, and northward from the Rio Grande to the Great Slave Lake in Canada, the country teemed with their numbers. They were massed in such crowds that as far as the eye could reach the land itself was invisible. At the time of the discovery of the North American continent, fully one-third of its whole area was covered with bison.

Sixty million of them packed the plains, yet the first one ever beheld by foreign eyes was a solitary captive, behind the bars of a cage.

In 1521, Cortez and his army, the first of the Spanish Conquistadores, entered America. They came by way of the Pacific, and, marching overland across the sunny Southwest, invaded the kingdom of the Aztecs. Here they found the riches of which all the earlier treasure-seekers had dreamed. The Aztec king, Montezuma, lived in a splendor that dazzled even Spanish eyes. His capital was filled with palaces and temples of surpassing beauty. In one of the magnificent squares of the city stood a menagerie, and here Cortez, the conqueror, stopped to admire a great "Rarity," a handsome bison. Cortez had no suspicion that only three hundred miles to the north, the "Rarity's" wild brothers smothered the land.

AMERICAN BISON IN WINTER

It was a man of a totally different type that first recorded the existence of the bison millions in America. He was also a Spaniard, a shipwrecked adventurer whose boats were washed ashore on the Gulf Coast, nine years after Cortez' triumphant invasion. He was a picturesque ne'er-do-well, named Alvar Nuñez Cabeza, better known in history by his nickname *Vaca Cabeza,* which means "Cattle Cabeza." He probably had his own good reasons for leaving Spain, for he burned his ships where they lay on the beach, in order to destroy any lingering temptation he might feel toward a homeward voyage. He brought no glory to Spain. He made no conquests. He was a vagabond, and his wanderings brought him into the region of what is now the State of Texas. His first glimpse of a roaming bison herd thrilled him, not so much because he found them handsome beasts and impressive in their numbers, but because he was half-starved. In the story of his drifting journey over the plains, he makes special mention of the excellence of bison meat, and the comforting warmth of bison hide.

This was no discovery. The Indians had always known the animal's worth as food and clothing. The settlers and pioneers of later years, making their way across the stretch of plains, drew on the fat, thick-coated herds for all their needs. Even in the East, where there were small bands of bison roaming near the protection of the forests, they supplied the wants of those who first ventured beyond the scant farms and towns of the newly-settled land. Fur-traders and explorers found them in the wilds of western New York, Pennsylvania, and even on the site of the city of Washington. Only three hundred years ago, the United States was largely unmapped, untamed land. Men lived on what they found, and all the early homesteaders and travelers had good reason to be grateful to the bison, for no other wild creature had so much to give. In fact, he gave so much that it ended in his destruction. The small bands grouped in the neighborhood of New York and in the plantation regions of Virginia, North and South Carolina and Georgia vanished quickly, but the millions on the prairies were untouched until the time of the opening of the West.

The American bison has always been called the "buffalo," and he is popularly known under this alias. The name properly belongs to the wild ox of the hot countries, Africa, India, the Philippines and Malaya.

The true buffalo is quite a different creature. He is larger and bonier than the bison, broad in the flanks, slate-gray in color, is virtually hairless, and has heavy horns growing in a bold, wide sweep. He has no hump.

A WATER BUFFALO

Strangers to the wild game of America, hearing of both "bison" and "buffalo," have often concluded that there were two kinds of wild cattle on our plains. Buffalo robes, buffalo trails, the city of Buffalo—even the ringing name of Buffalo Bill, the man who became famous because he never missed a shot while hunting on the gallop—have so firmly established the wrong name that it clings beyond changing.

The form of our American buffalo is familiar to all

of us from his portrait stamped on the face of our current five-cent piece. The narrow hindquarters, the high hump just behind the neck, the shock of thick hair enveloping his head and shoulders, running down over his chest and covering his forelegs like a pair of shaggy stockings—even the details of his close-curving horns and the beard below his chin, all show clearly in the relief on the coin.

The individual that posed for this portrait must have been a bull between ten and fifteen years old. The clue to a bison's age lies in his horns, which change in shape and surface and color from infancy on. Hunters and plainsmen of the West used to distinguish the animals by three names, according to the state of their horns, as yearlings, spike-bulls and stub-horns. The baby calf showed two tiny, blunt stumps above his ears, which bore not the slightest indication of their future form. In fact, the calf was altogether lacking in resemblance to his parents. Before his birth, his mother usually withdrew to the shelter of a nearby ravine, where she waited a few days until the wobbly-legged baby could amble into the herd under her protection. He looked strangely out of place among his grown-up relatives. Their big, dark bodies towered over him. He was little and scrawny. His hair was fuzzy and tow-colored, strongly tinged with red. He had neither mane nor beard, and hardly a hint of a hump. He looked more like a spindly-legged foundling of another breed than like an heir to bison blood.

However, by the time he was six months old, he began to take on the family resemblance. His reddish hair had dropped off and his new coat was coarse and dark. His body was fuller, and his legs thicker. His horns still showed no bison grace. They had grown from knobs to a three-inch length, but where was their curve? They jutted out perfectly straight, like two little knife handles stuck there at an angle. As long as his horns were in this awkward state, he carried the name of *yearling,* which applies to a calf of any breed.

It took several years for them to develop the curve. Very slowly they changed their rigid, stubborn lines, and began to sweep slightly in a shallow arc. During this period, he was promoted to the rank of *spike-bull* by the plainsmen. He kept this title for four years, while his horns increased their arc until they looked like the two branches of a candlestick. The curve at the base hardly showed, as it was hidden under the thatch of dark hair over his ears.

The young spike became a full-grown bison only when the tips of his horns finally began to curl inward. Then they were all that bison horns should be—smooth and black as polished iron, shining in the tangle of dark hair, strong enough for battle, graceful enough for beauty. The bull is in his prime when he enters his tenth year. It was at that age that he fought for first place in the herd. He kept his mastery for a few years only, and then began to age. His horns again underwent a change. They became thick and rough. They lost their sheen. They even changed in color. With digging and fighting and friction, they became split and scarred. The glossy tips had cracked and curled backward in clumsy rolls. Ever since his spike days, rings had been forming at the base of his horns. In his old age, these grew lumpy, gray with grit. For the last years of his life, the bull was known as a *stub-horn.*

The horns of the cow showed fewer changes. They were black and shiny, also, but more slender and shorter. They took on the curve more rapidly, and frequently one of the pair was "crumpled," or otherwise out of shape. Her horns were never as handsome as the bull's, nor did they become so sadly disfigured.

Yearlings, spikes, adults and stub-horns, all depended on the cows for leadership. If the mother started on a trek, the whole family set into motion. If she was content to linger for days in the same patch of plain, they lingered with her. The one strongest instinct of the buffalo was the bond of the herd. Parents, brothers and sisters, cousins and aunts and uncles clustered together in groups of fifty to a hundred, and these tight little family groups merged together into bands of thousands. Where they pastured, the herd looked like a black island on the prairie sea. Where they took to the trail, the herd looked like a twisting black ribbon dividing the plains. Seen one by one, bison are not black. The hair of the robe shades from tobacco brown over the body and hindquarters to hay-color over the hump and to deep brown in the shaggy mane and the rough mantle that covers the shoulders and forelegs. The thatch that runs around the eyes and horns and down the length of the nose is still darker, and the beard, which grows to a length of twenty inches in the sturdiest bulls, is quite black.

The hair on the narrow flanks is packed close to the hide, almost like wool. The hair on the fore-quarters is long and dense. In the bitter cold of the open winter, this frontal shield of fur was a tremendous advantage to the buffalo. Other cattle, caught in a storm, invariably turn their backs on the beating wind and drift helplessly wherever it forces them, but the buffalo faced bad weather with persistent obstinacy. No matter how strong the wind, no matter how fierce and cutting the blasts, or sharp the hail, or thick the blizzard, they

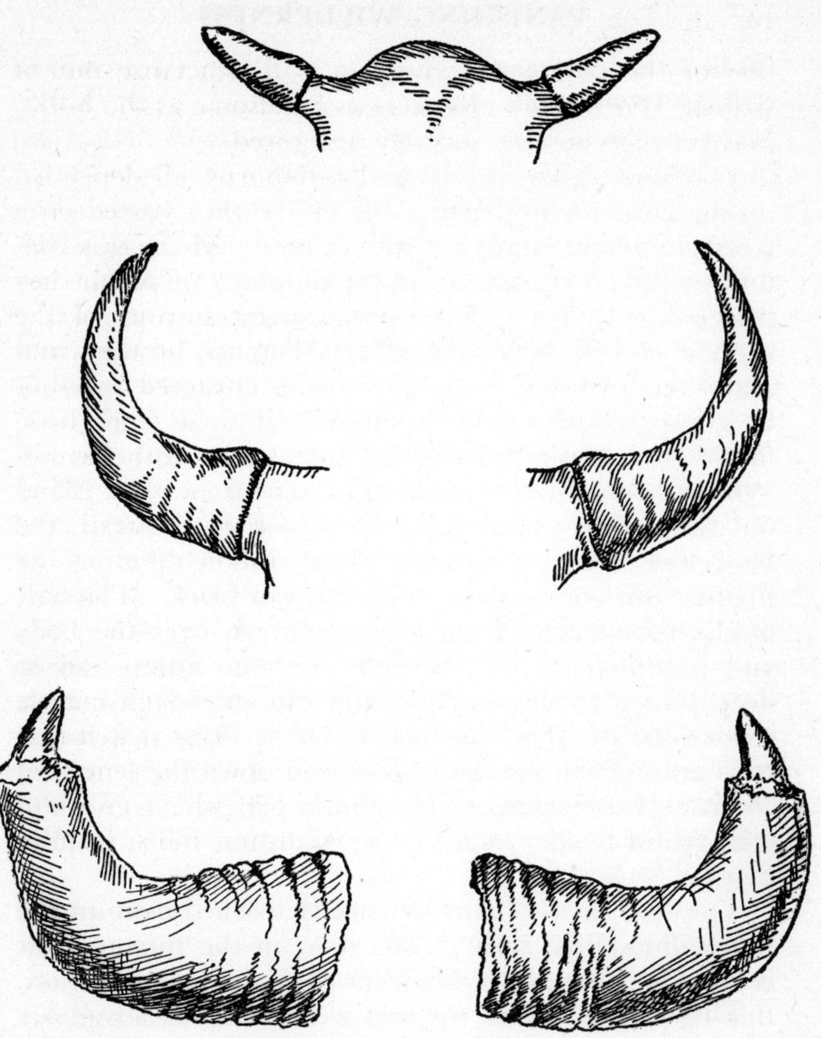

THE BISON'S HORNS ARE A CALENDAR OF HIS YEARS

Top: SPIKE-BULL
Center: ADULT BULL
Lower: STUB-HORN

pushed straight into it, head on, resisting the elements with all their might. Bison instinct drove them forward, and it was well that they were so warmly protected for the face-to-face conflict with the cold.

The herds made long journeys as the winter grew more severe, sometimes traveling two or three hundred miles before they came to a halt, but they did not migrate along fixed routes, nor did they head toward warm country. The huge extent of territory, all the way from Canada to the Rio Grande over which they spread, offered no great "summer resort" region. They did not need the warmth of the South. It was not the cold that determined their wanderings, but the search for food. They could stand the sweeping storms, especially if they made their way before the heavy winter toward the line of sloping hills and ravines. There they could take shelter when hard-pressed by the blizzard. Even where the snow lay thick, they could plow through safely. What they had to have was a continuous supply of grass, even if they had to paw through inches of crusted snow to reach it. They must not, however, be caught in the "wide, open spaces," where there was no dip in the land to break the force of the wind and the engulfing miles of snow-drifts.

So, every year, the herds deserted the broad plains which gave them such rich feeding in the summer and became so inhospitably shelterless in the winter.

The buffalo's home shifted with the seasons, and his appearance changed with them, as well. At times he was handsome; at times hideous. He began the winter, a lordly aristocrat. He faced the spring, a dreary scarecrow.

After the easy, rich feeding of the warm months the herds were fat and full of vigor. The big bulls reached a weight of two thousand pounds and over. The best of them stood six feet high at the hump. Their robes were thick, their horns glistened. Their beards added to their dignity. Black horns, black hoofs, black tassels tipping their tails, their massive bodies a blend of warm tan and brown, they deserved the title they have carried as "America's handsomest game."

But at the coming of spring, they had no right to such a compliment. When they returned from their migrations, they were a horde of unkempt, bedraggled tatterdemalions. The well-fed look was gone. The aristocrat of the winter was a caricature of his former self. He was bony. His luxuriant coat was in shreds. He was faded to a dingy mud-color. His hide was blotched with bare spots. Ropy strands of bleached hair hung from his matted mane, like icicles of wool. He was a sorry sight, the picture of poverty and woe. Yet there was nothing seriously wrong with him. He was merely beginning to go through the yearly change of coat.

The shedding of the hair was a slow process, lasting from late February to late September. Even while the new coat grew in, the old wisps clung in untidy tangles. Through the pleasant months of spring and summer, the buffalo grew uglier from day to day. He also grew more uncomfortable. Every time a flake of his old coat dropped off, a fresh swarm of mosquitoes and prairie pests fastened on his smarting hide. They fed on him voraciously. He itched from head to tail, and the biting rays of the hot prairie sun did not add to his comfort. He was thoroughly miserable, and he looked it.

Something had to be done to ease his torment. Wherever a tree offered its rough bark, the animals took turns rubbing against it. Here and there, a gritty boulder invited them to sandpaper their sides against its surface.

Early travelers in the West saw these trees and rocks polished to a gleam by generations of itching bison. In later days, when the first telegraph poles were set up on the plains, the beasts promptly wrecked them by the force of their scraping bodies. The exasperated linemen thought they would punish and outwit them by

HE LOOKED MISERABLE

driving sharp spikes into the new posts, but the iron points only added to bison satisfaction.

Trees and boulders were all too scarce on the plains, and the universal cure for the annual itching was the wallow. The plains were mutilated with numberless cavities in which the animals rolled and twisted and rubbed and shoved in frantic efforts to find relief. Tiny mud-puddles and depressions were broadened into bowls. On flat land, where there was no hole ready-made, they desperately dug their horns into the dusty soil and hollowed out their own excavations. One after another, the big bodies churned the spot deeper and deeper.

When the rain fell, the dust bath became a mud-bath, which was even more to their liking. The mud clots did not improve the bison's scarecrow appearance, but they were a wall of protection against the mosquitoes and the stinging sun.

Long after the bison had vanished from the plains, some of these wallows remained as ornaments in the waste. Where the moisture had gathered, rich green weeds sprang up and formed little garden spots, tiny oases in the burnt stretches of brittle prairie grass. They gladdened the eyes of the pioneers with their fresh, cool charm. They were known as "fairy rings" among the travelers, and at first the cheerful, miniature patches of green in the dry, caked country could not be accounted for. But plainsmen familiar with the buffalo's problems during the shedding season recognized them as wallows.

All this rubbing finally tore away the last shreds of lifeless hair, and the new coat was free to thicken and lengthen. By August, the bison began to resume their good looks, but they lost their good temper. This was the time of battles, of rivalry and of hard-fought butting contests. The herds grew restless. The family clans no longer stood in separate groups. They surged together in a close-packed mass. The mating season had come, and the old stolid calm was gone. Terrible combats took place. The young bulls, confident of their strength, tried out their mettle against their elders. There were vicious collisions and crashes as rivals banged their heads together in stubborn duels. Sometimes they used their horns, but most of the fights ended without bloodshed.

The older bulls struggled to keep their rank in the herd. Every group had its "boss" and his position was worth defending. It meant first chance at all the privileges of good grazing, first drink at the stream, first choice in mating, first rights at the wallows and rubbing places, and all the other privileges that might belong to the master of the clan.

The young bulls seldom won at their first tournaments, but they tried again the next year, and the next, until their strength prevailed or they resigned themselves to a humbler lot. Every August, new champions stepped into place to lord it over the others. The "boss" enjoyed his rule for about three years, and then had to yield to a stronger successor who challenged him to battle and defeated him.

An old bull who had been beaten took his disgrace much to heart. Once he had lost his supremacy, he lost his courage. The humiliated *stub-horn* drew back from his fellows. His pride deflated and his nerve gone, he straggled behind, hovering on the outer fringe of the herd. Sometimes, he removed himself entirely, and lived out the rest of his days in solitude. There were many such "lone bulls" wandering by themselves on the far prairies. They often came to a sad end, unable in their weakened and discouraged state to fight off a pack of prowling wolves.

During the mating, or "running" season, as the plainsmen called it, the herd was a surging, jostling mass, stamping and bellowing until the earth shook under their hoofs and the air echoed for miles with their roaring. The restlessness and fighting lasted for several weeks, but gradually the tumult died down. Again, they broke up into clusters of family groups, and devoted themselves to placid grazing.

The buffalo fed on whatever grasses he found, even on sage brush, when nothing better offered. His main support was the dry grama grass, which hardly looks like rich eating, but was more nourishing than the juicier herbage which he took as second choice. One thing he avoided—the "loco-weed," which the horses and cattle of the ranches ate with such sad results. The bison seemed to know that it was bad for him, and he escaped the fits of illness and frenzy that attacked these animals when they fed on it.

For several days the herd would remain at pasture, grazing during the early morning hours, and lying down to chew their cud and rest when the sun grew hot. They did their first chewing hastily and swallowed the food into the paunch, a sort of "first stomach," where it was stored. While they rested, they drew it back into their mouths again for more thorough mastication, and then swallowed it for final digestion, as all ruminants do.

Every two or three days, the buffalo drank. None of them ever wandered off in search of water by himself. It was their nature, under all circumstances, to move in a mass. When the leader became thirsty, she raised her head and sniffed the air this way and that, and started forward. At once, thousands of buffalo found themselves ready for a walk. They not only followed their leader—they stepped in her footprints. They marched in single file, each one planting his hoofs in the narrow trail, swerving not an inch to the right or left. Under their heavy weight, the path was worn into a shifting, uncomfortable groove, and the going became harder and harder. At last, some one enterprising buffalo, most likely a cow, with a mild degree of leader sense, took courage and ventured a little to one side of the beaten trail. Immediately those behind her followed her bold steps, and the single-file parade went on.

Streams are not frequent on the dusty plains, and often the herds tramped for miles before they came to water. The leader took no short-cuts over the hills, but wound around them, preferring a long trek to a climb. The trails were so deeply cut and so sensibly directed over the "easiest way" that they saved the pioneers in their covered wagons a great deal of trouble when they traveled through buffalo country. The prairies were

BISON ON THE TRAIL TO WATER

slashed with ready-made routes of gentle grading, clearly marked for their wheels, leading to the water beds.

The end of the water trek was not always a running stream, but even a muddy pool was better than nothing for bison thirst. At the drinking place, the "bosses" had first turn, and the others fought and pushed for the privilege of guzzling after them. After drinking, they all lay down to rest, but not by any means heedlessly. Every member of the herd pointed his head toward the wind, according to fixed and unchangeable bison custom.

They seemed to want the wind blowing straight into their wide, black nostrils, at any and all times.

A few hours of rest, a few days of grazing in the neighborhood of the water-bed, another drink—with the same squabbles for first place repeated—and the leader again took up the march, with the herd in single file at her hoofs, on the way to fresh pasture.

The burden of leadership fell on the cows, but the bulls assumed the duties of protection. The hungriest prairie wolf would not dare attack the grown-ups, but a helpless little yearling would be easy prey—if the wolves could only get at him. They often skulked along the fringe of the herd, approaching in packs at nightfall, watching for a stray calf. They waited patiently, sometimes for hours, licking their chops in anticipation of a tender dinner. But the bison seniors massed themselves in a defensive circle around their children, ready to give battle with their hoofs and horns if the wolves grew bold enough to attack. In the calving season, the cows remained in the hollow of the ravine until their babies were strong enough to wobble into the herd. The bulls ranged near by, and always answered a call for help.

The buffalo had other perils to meet, against which horns and hoofs and fierce fighting were of no avail. The winter blizzards sometimes engulfed thousands of them. Sometimes, whole herds were trapped in quicksand, or drowned while swimming across a river on their long journeys. Such catastrophes were due to their complete slavery to the herd instinct. Not even to save their lives could they free themselves from the impulse that made them cling together. Where the leader went, the mob followed, regardless of safety or danger. They might see hundreds of their fellows floundering in the bog, drowning in the river, their heavy, jostling bodies too close-packed to give any one a chance to make his way, yet the pushing army came surging on from behind. It is doubtful whether any animal that ever lived was so obsessed with the herd instinct.

The Indians counted heavily on this herd instinct of the buffalo in their hunts. They had need of their knowledge of all the animal's habits, for the buffalo was the mainstay of their existence. Every scrap of him, inside and out, served some purpose to the Redskins. They ate his meat fresh in the summer, and dried it for the winter or mixed it with berries and fat into a concentrated preparation called pemmican. They flung his hairy robe on the floor of their tepees and slept on it. They dressed the hide and stretched it over the framework of their wigwams and over the wooden skeletons of their boats. They cut it up into moccasins, skirts for their squaws, fringed trousers for themselves and their sons, all of which were sewed with taut buffalo-sinew as thread. They even used his hide in the making of their literature. While poets and historians in other lands wrote their works on parchment, on tablets, and on paper, the Indians recorded the history of their tribes and the prowess of their braves on the scraped skin of the buffalo, not in words, but in colored picture-writing.

They made musical instruments of his bones, particularly of one part of the shoulder blade. It was fastened to the end of a thong made of buffalo hide, and swung through the air until it "roared" almost as loudly as the animal himself. The hoofs were boiled into glue for the feathers that adorned their headdress and sped their arrows. They shaped his hollow, black horns into cups and spoons.

The Indians and the white hunters of the buffalo alike placed a high value on any buffalo robe that was unusual. Occasionally, there would be one with a blueish tinge, or a "black and tan,"—jet black except for pale markings on the nose, flanks and forelegs. The two finest were the "beaver" and the "buckskin," the one a shining brown, with the hair of a soft, silky texture (it was generally the hide of a cow), and the other a dull cream. Very, very rarely, perhaps not more often than once in two million, a pure white bison was brought down. When the Indians laid hands on such a treasure, they dedicated it in solemn ceremony to the Sun God.

They did most of their hunting in the fall, when the robes were at their best and the animals at their fattest. Once in a while, they went out in the dead of winter. When their scouts reported that a herd had been trapped in a snow-drift, they organized a snow-shoe hunt. The lithe Indians could skim over the crusted snow and come quite near to the floundering beasts, and shoot them at close range with their arrows.

One of the boldest and most successful methods was to drive the buffalo to their own destruction. The Indians knew every inch of the plains on which they lived, and where they came upon a herd grazing on a stretch of table land that ended abruptly at a precipice, they managed the hunt through strategy. One of their braves, dressed in a shaggy bison-robe, took up his stand at the brink of the cliff. The hunters, stealing up on the quiet herd, broke suddenly into wild whoops and yells, and shocked the animals into panic. They took care to frighten the leaders first. The beasts plunged and snorted in frenzy, and dashed in a mad stampede straight toward the decoy, hidden under his buffalo robe, at the edge of the precipice. Once they had started, there was no stopping, and they lunged on in a mass, crashing in hundreds over the cliff. The Indians slipped lightly down the mountain side to the base of the steep bank. There lay the bison herd, ready for their skinning-knives.

Though these wholesale hunts had gone on for untold years, though the herds perished by thousands in storms and streams and quicksands, still they increased so plentifully that they could have survived all such tolls. But the sixty millions came to a quick end. The United States was destined to change from a wilderness to a highly civilized country. The buffalo was destined to make way for that change. In the short span of a hundred years, a very brief period in history, the face of our prairies was transformed. The pioneers opened the West. They drove their covered wagons straight through bison country, actually on his tracks. Like the Indians, they were obliged to kill the buffalo for the warmth of his robe and for the sake of his meat. He kept them alive. But their forays did not exterminate the great herds. The buffalo retreated from the settlers' routes of travel, but their retreat was of no avail. A new menace overtook them.

All over the country, East and West, North and South, a demand for buffalo robes sprang up. They were warm, they were strong, they were easy to get. In about 1820, the prairies were invaded by bands of hunters, who pitched their camps in the neighborhood of the grazing herds and went seriously to work with their rifles. The old method of hunting the buffalo on horseback was abandoned in favor of the "still-hunt." A single marksman, lying flat on a ledge of table land, or hiding in a gulley, could pick off three times as many buffalo in a season as twenty men on horseback, if he knew enough to keep out of sight and to give the leader of the herd his first shot. It was the old story. If the leader did not move, the herd did not move. Once she was down, the rest were at the hunter's mercy. Bison were always stupid. Even their greatest admirers admit that. The ping of the bullet, the puff of smoke, the crash of their comrades' falling bodies as one after another dropped to the ground meant nothing to the herd. They were terrified, but it never occurred to them to run. They stood stock-still, waiting for their leader to rise and move off. They waited in vain, and one by one the hidden marksman picked them off. With luck and steady, hard work, he might bring down a hundred buffalo in less than an hour! If his rifle got too hot, he stopped long enough to plunge it into the snow to cool it. Then, back on the job!

The opening of the railroads in the West made shipment of the hides easy. The hunters worked harder than ever. The bison herds dwindled. The prairies, once black with their millions, were now white with their bleaching skeletons. A new trade sprang up. A new band of treasure-seekers invaded the West. Crews of teamsters gathered on the plains and raked up the heaps of bones, piled them high in their wagons and drove off to the waiting freight-cars. The country's farmers clamored for bison bones, for they made excellent fertilizer. Most of the prairies were deserted by the living animals, and swept clean of the skeletons of those that had perished.

Still, there were regions where the buffalo survived. Along the foothills of the Rockies and in the forested Athabasca district of Canada the wood bison roamed, smaller in size and darker in coat than their plains brothers, and grouped in lesser bands. The cramped spaces of the woods prevented their associating in huge numbers, and they had the protection of the forest further in their favor. All buffalo, both of the woods and the plains, could be surprisingly nimble-footed in spite of their heavy bulk. The biggest bull could scramble up a steep ledge safely, and often made his way, when sick or wounded, to some high cleft in the rocky wall, filling the opening with his great body and offering his horns to any wolf that might follow him to his retreat.

The wood bison suffered less from the hunters because they were inaccessible, but the herds of the plains had no refuge. Their only resort was their ancient tendency to stick together. Safety in numbers! From far and near, they drifted together, finally concentrating in one vast unit, known as the Universal Herd, in a last effort to combine forces, true to their instinct for companionship. But their effort failed.

The Union Pacific Railway was completed in 1869, and the Universal Herd was split into two sections, the Southern Herd centering in and around Kansas, the Northern ranging from western Nebraska and Wyoming to the shores of the Great Slave Lake. The division did not, of course, take place all at once. Small bands of buffalo roamed for a time right in the path of the railway. Often, the train had to stop because there were bison in the tracks, obstructing traffic. More than one passenger stepped out of his coach, gun in hand, and took a shot at a bison or two. The beasts snorted and stamped, and in their panic and confusion, some of those on the side climbed up and planted their hoofs on the very track-bed from which they were being driven. Finally, the dullards learned that they were not welcome and the herds wandered away to territory that was still wild. At the time of the break-up of the Universal Herd, the Southern Herd numbered about four million, and the Northern, half as many.

There were still enough buffalo left to satisfy the demand for buffalo robes, buffalo tongues and buffalo bones. Every fall saw the same hunting outfits embarking on the season's work. But the work became harder. A strange thing happened. The animals made a discovery! They somehow found that it was possible for an individual to move without waiting for the mass. The marksmen aimed first at the leaders, as they always had, but the bison no longer stood waiting for their leaders to rise. They were no longer a collection of statues, inviting their own destruction. The first shot sent them into panic. They ran!

They had at last learned the lesson of flight, but the lesson came too late. In four quick years the Southern Herd was wiped out. The Northern Herd lasted about ten years longer.

Far off, here and there in outlying lonely wastes, or in the seclusion of the hidden ravines, a few scattered bison still lingered. In pitifully small clusters, ten in one state, twenty-five in another, the last of them still clung together. There was even a tiny herd in western Dakota, numbering four!

The government of the United States, in response to the persuasion of a group of men who interested themselves in the fate and future of the American bison, decided to collect these tragic remnants of the once teeming herds and give them a safe refuge. The country was scoured by bison hunters, this time on a merciful errand. After a thorough search of the West, state by state, they reported that they had mustered eighty-five wild buffalo. The animals were sent to a preserve in Yellowstone Park.

Even in the postscript to their history, the buffalo met with reverses. Some of them wandered beyond the protected range, and fell to the rifles of casual hunters.

Some were overtaken by poachers, who ruthlessly ignored the boundaries of the Park and took what "luck" brought their way.

Stricter laws were passed, and several other reservations were established, and in order to restore the scant numbers, additional bison were bought by the government from ranchers who owned them in their private corrals. In Canada, the wild wood bison were likewise placed under the protection of the government, in a huge forest in the Athabasca region. Their census showed a count of five hundred and fifty.

Safe, at last, in these retreats the bison of the North American continent have multiplied, and their total now reaches the proud figure of twenty thousand.

The closest relative of the American buffalo is the wisent of Europe. He, too, has all but vanished, but he, too, has found his friends. There are twenty-seven wisents in England and ten in Germany, in a semi-wild state. They are privately owned and kept in parks on great estates. Poland has recently issued an order that all the wisents in the zoos of the country be purchased and sent to the Forest of Bialowieska, the scene of their ancestors' wild existence. They are being repatriated, restored to the woods and ways of their fathers.

2

THE HIPPOPOTAMUS

THE Roman emperor Octavian conquered Egypt, but he failed to bring back its most beautiful possession —the imperious Queen Cleopatra. So instead, as a symbol of his conquest, he brought back and paraded through the crowded streets of Rome, a live hippopotamus!

This large and unattractive animal, whose home is now practically limited to the portion of Africa between the Nile cataracts and the Limpopo River, was then plentiful all over the huge and unsettled Dark Continent.

After the strange creature had been carried through the streets of Rome, it occurred to some one that they had here a completely new animal to use in their favorite sport—the gladiatorial contest.

The Roman gladiators were a specially trained body of men, trained in schools for fighters. They were divided into different classes according to the ways they fought. Some fought men and some fought animals; some fought by wrestling only, but most of them fought with dangerous weapons. The majority fought to kill, but there was one group, the Pægniarii, who used harmless weapons and only indulged in a sham fight. The Retiarius combat consisted in capturing the opponent with a cast net, after which he was killed with a trident. One class fought on horseback and one from chariots.

There was even a class who tried to lasso their antagonists.

This sport was so popular that occasionally the great nobles themselves entered the arena, and there are even records of athletic Roman women who were cheered on by the crowd! If a gladiator was wounded, the decision as to his fate was left to the audience. If the audience decided in favor of mercy, they waved handkerchiefs; if they were in favor of death, they turned their thumbs downward.

Against the agile gladiators with their sharp weapons, animals had but small chance, and the new animal, the hippopotamus, had even less chance than most. In the first place, he weighs an average of three tons, is about fourteen feet long and ten around, and, in spite of his immense length, is very much underslung, only about five feet from the ground at the shoulder. His head, flat on top with small, high-set eyes, little ears and a huge muzzle, sags very close to the ground. Its weight seems burdensome even to the hippo himself, for he is very apt to rest his chin on something, generally another hippo.

His enormous body rests so heavily on the short, wide-set legs that the animal's footprints are pressed deep into the soil, and make two deep furrows with a high ridge in between—a ridge often scraped on top by his stomach. This, with the mark of the four toes, each enclosed in its own hoof, makes a track well known to African natives and hunters. The track is worn deeper and deeper each year in the hippo's own districts by the animal's passage back and forth from his river or lake home to his feeding grounds. Travelers, man and animal, are often glad to find this ready-made path through the swamps and marshes.

But in spite of his awkwardness on land, once in the water, where he spends most of his life, he is a buoyant creature. He swims, floats and dives with ease and speed. More usually than bothering to dive to the bottom of his river, he just lets himself sink, and once arrived there, trots nonchalantly along the river bed with a foot or two of water above him. He can stay under water

NO ONE LOVES THE HIPPOPOTAMUS

this way from five to eight minutes and perhaps longer. He naturally breathes very slowly and the places where water might enter to choke him can be made water-tight by means of muscles that close flaps over his ears and nostrils. His infolded lips shut compactly.

When he swims along, low in the water, just the flat top of his head exposed, he is next to invisible. His grayish hide blends with the color of the muddy water and the river weeds.

Because of his ability to run along under water, or to swim so low that he is hard to see, the natives have devised a special method of keeping track of him in the hunt. The Shilluks, one of the Nile tribes, hunt him with the harpoon. They know that one thrust usually does not kill this animal whose hide is two inches thick, but only wounds it, and the first thing the wounded animal will do is to swim off as rapidly as possible, or to go under. So, in order not to lose sight of the hippo, the Shilluks fasten to the shaft of their harpoon a long rope of twisted grass, to which is attached a spray of faggots. When the hippo, with the harpoon sticking in him, disappears from view, the twigs remain floating on the surface of the water, and thus enable the hunter to follow the beast and attack him again, or take him when he is exhausted, for he cannot swim fast for a great length of time.

The Africans have always been great hunters and the hippopotamus has always been hunted. The ancient Egyptians hunted him in boats, for in those days he and the other river terror, the crocodile, both frequented the Nile.

The hippo, except for freak attacks, avoids human encounter when he can. His scent is very keen and if he suspects the presence of man, he will even abstain from his usual puffings and wheezings, or his harsh, loud bellow.

This animal was, and still is, hunted not so much for sport as for useful purposes. His plentiful flesh supplies a welcome variation in the diet of many native tribes who are otherwise almost entirely dependent on fish. His tough, warty hide makes good polishing wheels and whips. This loosely hanging hide, when the animal emerges from the water, sometimes exudes drops of an oily reddish liquid that acts as a natural lubricant. Deprived of this oil, the hairless skin rapidly dries and behaves as if it were badly chapped. When a hippo is brought on the long voyage from its African home to an American or European zoo, the keeper in charge of it on the trip watches for this drying, and if it begins, he rubs the animal with oil to prevent painful sores from developing.

But the hide, so sensitive in life, is extraordinarily tough and impenetrable when dried. Some of the most warlike tribes have long used it for their shields. It makes particularly strong whips and is so generally used for them that in several dialects there is no word for whip. When the Swahili wants his whip, he says, "Bring me the hippopotamus"; when the natives of some of the lake tribes around Tanganyika or Victoria want a whip, they ask for a *tomobo*—the native word for hippo.

There is another very important part of the hippopotamus. This is the content of his massive jaws—his fearful array of teeth. His mouth turns way up at the corners as if he were smiling, but any suggestion of a smile vanishes when the mouth opens in a three-foot gape, disclosing the formidable array of tusks, spikes and enormous grinding teeth.

Up to the end of the nineteenth century, the tusk-like canine teeth of the hippo were very popular with the manufacturers of false teeth. They are quite a pure white, and keep their color better than the ivory of elephant tusks.

Among some of the West African tribes, these teeth play a prominent part in treatment of illness. The natives grind them hollow and fill them with medicine. Then, instead of swallowing the medicine, as one might expect, the patient ties the filled tooth to a string and wears it around his neck! This is supposed to drive away the ailment. If the patient is suffering from snake bite, the tooth must be held in the hand.

More important still is the rôle of the hippo tooth as money. It corresponds to small change, and with it can be bought pieces of cloth, food, or other less valuable things. It can even compose part of the purchase price of a wife! It and the cowrie shell are low currency. They cannot be exchanged for cattle, because cattle themselves represent high currency. Cattle and one especially rare shell may be exchanged for really expensive goods!

THE HIPPO OPENS HIS MOUTH

Hunting the hippo entails considerable skill on the part of the hunter, for the hippo has an uncertain temper. If food is plentiful and they are undisturbed, they lead a sluggish life, moving little away from their chosen home for the twenty-five years or so of their life.

Resting in the sunny shallows, they utilize their friends' thick, spongy hides as cushions. The whole small, fat persons of the young hippos are used as couches, and no matter how many times they extract themselves from beneath their elders, the elders placidly and forcefully reseat themselves. Over the resting hippopotami walk various birds, as if they were walking on an island, and attached to the hippo's person in repose there are generally a few leeches. These creatures, so unwelcome to other animals because of their blood-sucking habit, apparently fail to disturb their large host.

Sometimes the hippo lives in groups and sometimes singly. The old males, usually bad-tempered, are very often hermits and occasionally choose for their solitary dwelling a most comically small pool of water in some lonely marsh. Among these old males there spring up "rogue" hippos. Like rogue elephants, these are vicious old beasts, known, hated and hunted by the natives.

Most of their wandering is done at night. During the daytime they rest. The cow hippo plays with her ungainly infant in a clumsy way, and the adult hippos indulge in fumbling games, pushing and falling about. From the heap of hippo there may suddenly emerge the fiercely combating forms of two old bulls.

The hippopotamus and the crocodile have apparently come to a truce. Few fights have ever been witnessed between them. They do not seem anxious to test each other's prowess.

The object most apt to undergo attack from the hippo is a canoe. It frequently happens that a hippo, unprovoked as far as any one can see, will rear out of the water, and, plunging and snorting, advance upon a canoe, upset it and bite at it and its occupants viciously with those terrible, armed jaws. Recently automobiles have been charged at in this same unreasonable and ferocious way. As for animals, such as pack mules, who enter the river, no one can predict what the hippopotamus may do to them. These attacks in the water are particularly savage and swift. The ancient Romans realized fully that this heavy, lethargic animal could be aroused to a fierce attack, for their arena performances were varied by goading the hippo into a rage so that, for the sake of variety, instead of the hippo, a few gladiators were killed!

It is only fair to the hippo tribe to mention that this large, unpleasant beast has a relative of much more peaceful and pig-like temperament. This is the pygmy hippopotamus, known as the Liberian hippo. He is not nearly as large as his huge relatives; his legs are longer; he has not such an alarming array of teeth. He is a solitary and peaceful creature living in swampy forests and rooting around for his food like a pig.

More persistently harmful than the hippo's sudden attacks are his inroads on the gardens of the natives. In spite of his fierce jaws, the hippo is a strict vegetarian. His chief diet consists of the water plants, reeds and grasses of his home in lakes and rivers. On his shore-going trips, conducted chiefly at night, he shows a decided liking for garden produce. No native's garden patch is safe from his intrusion. His stomach can accommodate three or four bushels, so it is quite a simple feat for him to wreck a garden in one or two good meals. What he fails to mow down like a scythe, he uproots with his two-foot canine teeth. It is largely due to his inroads on crops that his stamping grounds have become so limited and that he is gradually being pushed back farther and farther into the swampy river wildernesses.

The natives are on the watch for him, kindling all-night bonfires to frighten him away. They are very apt to plant their gardens, especially sweet potato patches, down near the edge of a river, and they have invented a special barricade against this night thief. This defense consists of ropes slung between trees or poles. At intervals of about twenty feet from each other are suspended

THEY PLACIDLY SIT ON THEIR YOUNG

stout poles. The raiding hippo hits the rope; the poles swing out, and he is alarmed and beats a retreat.

Sometimes the natives of a village miles away from any body of water will wake up one morning to find small pieces of a sun-dried, little-known plant in front of their huts. They will finally recognize this plant as one that only grows in the water. Then they know that a hippo has passed through the village in the night—a hippo who, when he reared his bulky body up out of its river home, brought with him unconsciously on his back some of the small plants floating on the surface of the water. On a small and very ancient turquoise-blue statuette of a hippo, found in an ancient tomb, some long-ago artist has painted one of these small water plants!

Water plants are not the only things hippopotami carry on their backs. About the only commendable trait of this creature is the mother hippo's protection of her young. The small hippo is born on shore, and in spite of the fact that at five years it is full-grown, it cannot take care of itself immediately after birth, as, for instance, a newly born fish can. So the cow hippo takes it with her into the water, and there she nurses it and carries it around on her back, protecting it valiantly against the attacks of the ugly old males until it can fend for itself. The first hippopotamus to visit New York was one purchased in London for thirty thousand dollars, and brought over in 1853. This animal's mother was wounded during its capture, and for three days the wounded creature followed down the Nile after the boat carrying away its child.

It is a very difficult and expensive task to collect and transport alive such huge, unwieldy beasts, but worth it, for the animal's haunts are getting more and more restricted as settlement and exploration increase. The zoo specimen who stands unenthusiastically back to you, muzzling and munching his hay, may soon be a representative of a vanishing race.

3

THE GIRAFFE

"The stature of a camel
The skin of a panther
The head of a stag
The horns of a fawn
The hoofs of an ox
The back of a cock
The tail of a bird"

THIS is not a conundrum. It is a description of a giraffe, set down about a thousand years ago by the inexact hand of a Persian writer after his return from Egypt. For a long time Egypt had been the center of civilization. She was a ruling power in the world, and ambassadors from the several mighty empires of the Orient came frequently to her court on missions of friendship and trade.

When they returned to their monarchs, they had strange stories to tell of what they had seen in the land of the Nile. Merciless stretches of sand. Fresh, green oases. Exotic luxuries. The snow of ivory and the jet of ebony. Mounds of gems and masses of gold. Swarms of black slaves. Learned priests who knew the sciences and the arts. Mighty tombs and temples, standing firm in the shifting sand. Their walls were triangular in outline. It was odd how the Egyptians loved the triangle. It appeared everywhere in their architecture. It was their favorite pattern.

Strangest of all, the travelers had even seen a beast in this land of marvels with a triangular body! Its neck rose in the air like a tilted tower. Its back sloped toward its haunches like the slanting line of one of the pyramid tombs.

As Africa is the only country in the world where giraffes exist, strangers saw them there for the first time. Their surprise was unbounded. When they tried to describe them to their friends at home, they were at a loss for words. They could only repeat that the giraffe was unlike any other beast they had ever seen. When their listeners asked how the astonishing creature was called, the travelers hesitated. They had to make up a name for it in their own tongues. The Persians called it the "camel-bull-panther." The Romans invented the word "camelopardalis." The Arabs gave it the name "zaraf," which has been translated in two different ways. Some define it as meaning "one who walks swiftly." Others say it means "creature of grace." In English, he used to be called the *camelopard,* but we have finally adopted *giraffe,* from the Arabic.

We can have no idea of the utter amazement with which people regarded the giraffe, wherever they saw him. It was the custom in ancient days for kings to exchange presents. Ships sailed home from Egypt, laden with treasures, with carvings and rich fabrics, with fruits and incense. And sometimes, the holds were filled with living freight. Rare birds screeched, and wild beasts snorted in their cages on the slow voyage across the seas. A wild lion or a bulky rhinoceros seems to us rather a doubtful expression of goodwill from one monarch to another, but the kings who received them were highly pleased with their novelty.

Of all the gifts that Egypt was in the habit of sending, none caused more delight than the giraffe. It was no easy matter to transport the long-necked, nervous prisoner in the clumsy ships of olden days. He had to be

GIRAFFES IN FLIGHT

carefully tended and fed. Swarthy Nubian slaves were torn from their homes and sent as part of the present. They were familiar with his needs and his timid nature. Special foods were prepared, packed in bales and kept fresh during the months of travel. They carried grain, beans and millet cakes for him, to substitute for the moist, juicy leaves of the acacia trees on which he had browsed on the plains. At the base of his fantastic neck hung a rope of massive ornaments, and from his muzzle swung tough leather reins with brilliant tassels.

Thus attended and bedecked, the spotted exile traveled eastward, to Persia, to India, and some were even sent to the far distant land of China. He was a source of wonder, wherever he went. The Chinese opened their narrow eyes in a stare of bewilderment. Persians stroked their beards in fascinated surprise. When the kings and princes had gazed their fill, the grotesque beast was led through the streets and villages. Men stretched their necks for a glimpse of his head. His picture was drawn and painted, stamped on prints, and woven into tapestries. Learned historians wrote about him. Poets exalted him. But no one knew anything of his life on the African plains. He was a curiosity, not a creature of the wild.

There was a time when people in Egypt had never heard of the giraffe. It is true, his home is in Africa, but only south of the Sahara. About four thousand years ago, one of Egypt's ambitious queens sent a fleet of ships to the coast lying south of her own borders on a voyage of discovery, and it was then that the existence of the giraffe was revealed to the world beyond the plains. When the Queen's expedition returned with its news, this ruler was not content with descriptions of the animals. She ordered her fleets to return to the land of Punt, where they had been seen, to bring back several of the animals. It was not primarily for the giraffes that her navy invaded the land of Punt, but for conquest, and when the tribes to the south were enslaved, they had to pay a heavy tribute. Egypt demanded treasure from her new subjects, and camels, which, as beasts of burden, were the equivalent of wealth in that land of sand, but she also levied a toll of two giraffes yearly. When her empire spread, it included giraffe country, and she could get them for shipment abroad within her own confines.

Europe saw her first giraffe in 46 B.C. when Cæsar showed them in his triumphal procession after his Egyptian campaign. The dignified Romans lost their dignity at sight of the weird beast. No matter how often they saw him, they never got used to him. Whenever a giraffe was promised for view, slaves, senators and philosophers mingled in a mob, pushing for a glimpse.

After the collapse of the Roman Empire, Europe forgot all about the giraffe. Dull years followed the fall of Rome. Gone were the exciting circus games, the triumphal processions, the hectic days of foreign conquest. For about a thousand years, no one in the western world saw a giraffe or knew of his existence.

Then, in the thirteenth century, he was discovered all over again. Frederick II, King of the Two Sicilies, was a man with a hobby. He was intensely interested in animals. He built a menagerie on his great estate. He wanted a giraffe above everything. He offered the Sultan of Egypt a white bear in exchange for a giraffe. He did not say which of the two kinds he preferred to have. He merely asked for a giraffe. When it arrived, Frederick was the envy of all his noble friends, and letter after letter was despatched from Italy to the Arabs in Egypt, who have always been the only traders in giraffes, and even to-day ship them to all parts of the world.

Paris and London saw their first giraffes three hundred years later. In about 1826 the Pasha of Egypt sent one with his compliments to the King of France, and a year later he bestowed the same honor on England. They caused as great a sensation in these sophisticated cities in the nineteenth century as they had in the ancient capitals of the Orient. The English restrained their wonder somewhat, but the French quite lost their heads with enthusiasm and surprise.

Our own country received the giraffe with less of a shock. We were prepared for his oddity through books and pictures. The first living giraffes that landed in America arrived about sixty years ago. By that time, popular interest in animals was not satisfied with astonished gaping. When men saw anything that was new to them, they were not content to exclaim about it and let it go at that. They asked questions. They wanted to know, not merely how wild creatures looked, but how they lived. Still, even then, there were many whose amazement outweighed their interest. There is a famous story of a man who literally refused to believe his eyes the first time he saw a giraffe. He stared, and stared, and finally gasped out, "Why, there *ain't* no such animal!"

So it has always been when the "tallest of mammals" has been seen as a spectacle, away from his African home.

A herd of giraffes, standing among a cluster of low-topped spreading trees is not bizarre—but beautiful. As they race over the plains, rocking their long necks in rhythmic balance, they are strikingly graceful. They are masters of the art of running, and masters of the art of standing still. They flock together in small herds, numbering between six and twenty. Sometimes an old bull withdraws from the herd, and goes about alone, forgetting the companionable days of his youth. No one has ever seen any sign of discord or battle among giraffes, but they may have some way of their own of indicating that one of the group is unwelcome, or it may be that the individual himself loses interest in the company of his fellows. Several cows and their babies and a young bull or two usually make up the herd.

Giraffes manage to live on the thin edge of the forest, here and there, but usually they are found on the plains where the mimosa trees and one particular kind of acacia, a thorn-tree, grow. They owe their existence to these

LEAF-PATTERNED GIRAFFE

trees. Among them, the bold pattern of the giraffe's hide becomes a cloak of invisibility. The russet and white of their bodies is lost against the rippling pattern of the moving leaves and branches. Their long legs are indistinguishable, at a distance, from the tree trunks. Their spotted bodies take on the confused blur of the foliage. Their long necks vanish among the leafy branches.

Small wonder that giraffes will stand for hours among the hospitable mimosas and acacias that interrupt the far reaches of the burning African steppes. They get their food from the tender leaves and shoots, their drink from the juice of the moist foliage, shade from the sun, protection from discovery, and even more. Their necks will reach right over the roof of the trees, and while they themselves are perfectly concealed, they can spy far across the plains. They cannot live in a treeless region, and are therefore not found in the desert.

If you should go in search of giraffes, you must travel south of the Sahara. The route of your journey depends on whether you are looking for the "reticulated" or the "common" giraffe. Most people look blank when they are asked which of the two they consider the more beautiful. They are alike in size, in form, and in color, but not in the design of their spots. Both are a medley of varying tones of brown and white. The easiest way to distinguish the two varieties is by remembering the simple words "net," and "leaf."

The giraffes in East Africa (in Abyssinia, Somaliland and the northern part of Kenya Colony) are marked in a somewhat geometric pattern. Broad white lines checker the russet of their hides into distorted squares, triangles and polygons of varying sizes and outlines. On the more slender parts of the body, the russet spots are smaller and fade into lighter tones, even into daubs of delicate fawn. The mottling runs all over the body, from the muzzle to the long tassel of hair at the end of his tail. All animals, whatever their species, have individual differences. Just as no two people look alike, no two animals look alike, and among giraffes, the color tones and arrangement of spots vary infinitely. But the distinct pattern is unmistakable. The "reticulated" giraffe looks like a rust-colored animal, with a coarse white net thrown over him—hence his name, "reticulated," which your dictionary will tell you means "covered with network." He is also known as the Northern giraffe.

Elsewhere in Africa, as long as you keep away from the desert and the moist jungle, you will find the common or Southern giraffe. His hide looks like a white robe strewn with giant sienna-colored leaves, irregular in shape and size and shading. He, too, is dappled from muzzle to the end of his tail, but his lower legs are often quite free from spots.

The giraffe is beyond dispute the "tallest of mammals." He holds this lofty title on two grounds, the length of his legs, as well as the length of his neck. Yardstick measurements are facts, of course, but they are not picturesque enough to enable one to realize how very high in the air this creature carries his head. Six feet is a good height for a man. Nineteen feet is a good height for a giraffe. The bulls average from sixteen to eighteen, and the cows are about a foot shorter. A baby giraffe is five feet high when he is born.

Three acrobats, standing "pyramid fashion," as they often do in the final stunt of their vaudeville act, would still be several inches short of an eighteen-foot giraffe.

If a giraffe happened to pass your house in a parade, he could easily peer into your second story window.

It is a relief to find that in one respect, at least, the giraffe's neck is perfectly normal. It contains only seven vertebræ, the number usual in mammals, the very same number that you yourself have. The giraffe's neck is

A HERD AMONG THE TREES

long only because each vertebra is of such great proportions.

He is not nearly so slenderly built as he appears. In fact, he is decidedly heavy, weighing well over half a ton, but the tapering line to his haunches gives a deceptive impression of slightness. The steeply sloping line of his back is not entirely due to the difference between the fore and hind legs, which does not exceed seven inches, but to a combination of the longer front legs and the bony structure of the shoulders, developed to support the weight of the neck.

Most mammals can be classed in family groups, according to their body structure, their teeth, their skull formation, and other characteristics. The giraffe is the

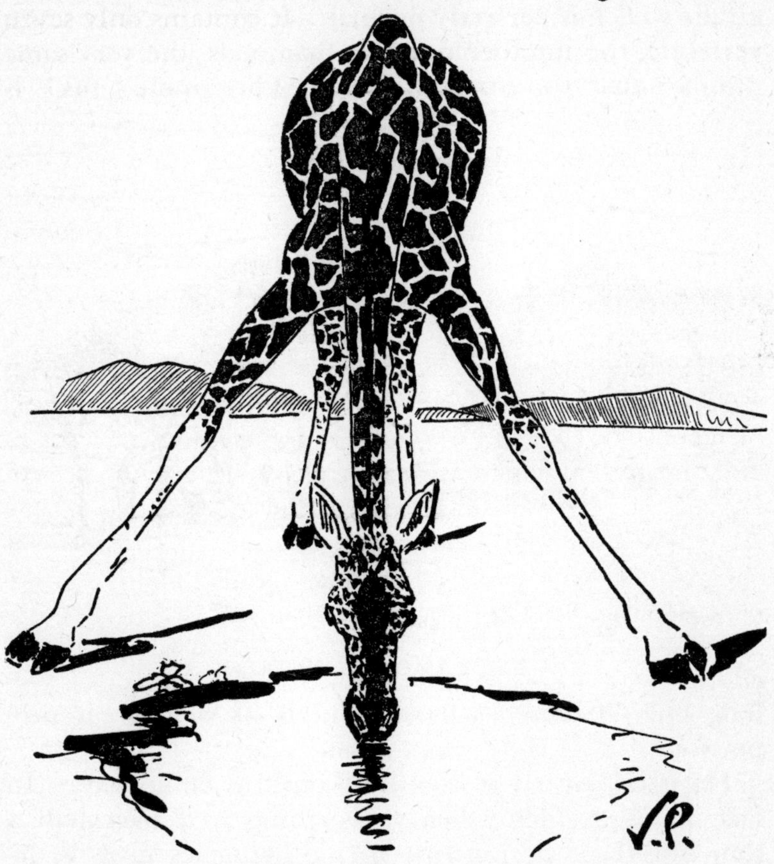

RETICULATED GIRAFFE

head of his own family, and an exclusive family it is. He has only one relative, very little known and very unlike him in outward appearance—the okapi.

Small wonder that the Persians resorted to the combination "camel-bull-panther" when they tried to think up a name for him. His spotted hide suggested the leopard, and the height of his body and the callosities, or pads, on his knees suggested the camel. Of the three, he most resembles the bull, or ox, although they are not of the same family. Like the ox, the giraffe has cloven hoofs, he is a cud-chewer, and his teeth are of the same order. He even has horns, though they are called horns only by courtesy, or for want of a better term.

Early artists who found the giraffe an interesting subject for their pencils, often omitted the horns entirely. Perhaps it is unfair to expect people to know much about the giraffe's horns. After all, they are pretty high up in the air.

"True horns" are essentially a form of skin. Our own skin is composed of two layers, the inner, or *dermis,* and the outer, which we call the *epidermis,* or cuticle. In various degrees of hardness and toughness, cuticle develops into hoofs, claws and nails. The "shell" of the tortoise is a form of cuticle. The "plate" of the armadillo is another form. Our own fingernails are likewise a modification of cuticle. When it appears in the form of weapons on the heads of animals, we call it *horn.* "True horn" grows as a hollow sheath over a bony core, as it does on all of the oxen, or Bovidæ.

But the growths on the giraffe's head are quite different. They are not hollow horn. They are bone. Furthermore, they are not bare, like the horns of cows, sheep and goats. The giraffe's horns are four to six inches long and covered with hairy hide, russet-colored and sometimes even mottled like the rest of his coat. Tufts of hair grow thickly around the base and a black band of smooth hair adorns their blunt tips. In a young giraffe, the bony knobs are loosely attached to the skull, and as he grows older they become firmly united to it. Old giraffes grow "bald" at the base of the horn.

Between the two horns, and a short distance below them, a third protuberance of bone rises on the skull. It is more like a bump than anything else. Both cows and bulls have these three bony growths.

There is a so-called "five-horned" giraffe in Uganda, west of the Congo. They are not very numerous. The two additional horns are quite short, and rise behind the ears. In size, this five-horned giraffe is somewhat larger than either the reticulated or the common, and the brown of his hide is liver-colored. His existence was not known until recently.

Horns do service to animals as weapons, but the giraffe never uses his in fighting. In fact, he never fights. He richly deserves the graceful name by which he is so frequently called—"the gentle giraffe." He is the most placid and peaceful of creatures. He is serenity itself, and even the most hardened sportsman feels a tinge of regret at hunting him. Giraffes are never attacked by other animals, though once in a great while a starving lion may try to bring one down, when he is balked of easier prey. When he is caught in a struggle, his only resource is the terrific force of his kick. His legs are powerful, and his hoofs heavy and large. The only trouble is that he cannot manage to aim with any exactness, and only accident or carelessness on the part of his attacker enables him to deliver a blow. A cow giraffe was once seen trying to fight off a leopard which had sprung at her calf. She could only fling her hoofs with desperate force at the enemy, but in the uncertainty of her aim, she struck the calf and broke its back.

With other animals he associates on the friendliest possible terms. Zebra, antelope and even the ostrich gather around the giraffe without fear. He may be seen at a water-hole in company with them. There he seems to be particularly welcome, for the beasts that are usually so afraid to lower their heads to drink at dark, do not hesitate nearly as long when there are giraffes at hand. One tall guardsman keeps watch while his fellows quench their thirst, and somehow a feeling of confidence spreads through the whole group at the pool.

Giraffes drink but seldom. They can go for weeks and even months without water, apparently without discomfort. Their favorite food is the foliage of the mimosa tree and the kameel-dorn, a thorn tree. They rarely graze. Their heads are constantly among the tree tops, on which they feed during the morning and late afternoon. The giraffe has a long, slender muzzle which can reach in among the crowded tiny shoots of the branches, and better yet, he has a foot and a half of tongue to help him reach the choicest leaves. His tongue is extensile and prehensile—that is, it has stretching power and clutching power. He does not bite at the leaves, but curls his tongue around them and draws them toward his mouth. His strong, mobile lips wait to grasp the leaves and his teeth to cut them. A growth of tough, black mustachios running all around his muzzle protects him from scratches against the bark and thorns of the trees. His nostrils are provided with muscles that enable him to close them against the pricking irritation of the rougher twigs and thorns.

With his head overtopping the trees, the giraffe spots every movement on the plains. His eyes, which are extremely beautiful, large and limpid and dark, are also extremely far-seeing. He is quick to take alarm, and his wariness is the despair of hunters. Africans have always greedily pursued him, though it took the keenest strategy to bring one down. Their spears were of no use at all, as they could never get anywhere within throwing distance. The only thing they could do was to lie in hiding for days and days near a patch of trees where a herd might come to browse, in the hope that a poisoned arrow might lay one animal low.

The Africans took a great deal of trouble in hunting them because they were trophies well worth getting. For one thing, there was the tail! Since very early days the giraffe's tail has been regarded as a thing of beauty. At the coronation of the boy king, Tutankhamen, when long lines of slaves filed before his palace bearing rare treasures from all parts of his kingdom, one of the most highly prized gifts of homage was the tail of a giraffe. To this day, value attaches to it. Men weave the black hairs into arm bands. Girls string their trinkets on them and wear them as proudly as if they were jewels. Even unadorned, a few of the hairs looped into a necklace is something to display, especially among the women of the eastern Bantus. As far as its value can be measured in money, a giraffe's tail is worth as much as twenty-five dollars.

The hide brings about the same price, not by reason of its beauty, but because of its toughness. It is an inch thick, and extremely durable, and makes sandals and whips that outlast any others, even those of rhinoceros hide. The native hunters feast on giraffe meat, lick their lips over the delicacy of the marrow of the roasted bones, and use every part of the animal. The tendons of the leg muscles make good sewing thread for their leather, and the Arabs in Abyssinia even use them as strings for musical instruments.

With the introduction of firearms into Africa, the hunting of giraffes became much easier, but it still demanded the highest skill and patience. Though most of the herd might be dosing or occupied in browsing, they would suddenly wake to life and dash off, apparently without reason. But they had had their warning mysteriously. In every herd, there is one who acts as sentinel. He stands a little toward the open, while the rest stand calm and serene with their heads lost in the cool foliage of the trees. It is not necessary for each one to be on the lookout. They need not even watch for the signal for flight. At the faintest sign of danger, the sentinel's pale, creamy ears point upward. He switches his tail. Alarm sweeps through the herd. Though they are such heavy animals, they skim over the plains with incredible lightness. Their tails twisted over their backs, their necks rocking back and forth in angular, easy rhythm, they gallop in swinging lines, a streak across the landscape. They can go at a speed of thirty miles an hour.

In swift motion, the giraffe loses his awkwardness. His walking gait is peculiar and clumsy. He advances both legs on the right side, then both on the left side, which makes his walk a series of jerks. But in flight, his body does not twist from side to side.

It is not only in moments of danger that giraffes profit from their herd habits. One of the most interesting features of their herd organization is the care and discipline of the young calves, who have not yet learned the lesson of moving in response to the sentinel's restlessness. They are watched by one of the cows, and the other mothers pay little attention to them. It is like some kindergarten class. The headstrong baby giraffes play tag around the trees, and cut capers, and as long as they stay within range of the group, the "teacher" allows them to frolic to their hearts' content. But whenever one of them strays a bit too far, she goes after him and shoos him back where he belongs, in the vicinity of his playmates. She takes her responsibility seriously, and watches over the children of her friends in the herd as well as over her own.

Giraffes usually sleep standing, with their heads nestling in the treetops. Occasionally, one of the younger ones may take his rest on the ground, but it is not their general habit.

Their drinking needs are not easily satisfied. Eating and sleeping, their long necks are a help, rather than a hindrance, but when it comes to stretching them down to the water, they have a problem to meet. It takes a giraffe twenty minutes and a lot of exertion to get a drink. Before he can get his mouth within reach of the water, his head has to make a descent of seventeen feet or so. His neck is not flexible. He can only manage by straddling his legs far apart to a stretch of three yards. Then, by a series of jerky contortions, such as only a giraffe could achieve, he manœuvres his head downward. The effort involves bending the foreknees and a complete readjustment of balance. His weight has to be thrown on his hind legs. Slowly, laboriously, the long neck descends. His mouth reaches the pool at last.

For a few minutes, he gives himself up to the long-delayed pleasure of drinking. He has earned his reward, but he does not take it all at once. After a few swallows, he raises his neck upward again, feeling his way into the altitude with nervous caution. Suddenly, in one last jerk, he is in place again, neck, legs, and all. For a little while, he stands at ease, recovering and catching his breath after the strain. Then he goes through the performance all over again. The second effort is no easier than the first. He sometimes has the patience to do it all a third time, though practice does not limber him up in the least. Of the twenty minutes he spends

YOUNG GIRAFFE AT REST

taking a drink, about five are taken up in swallowing the water. The rest of the time goes to getting himself into position and back again.

He may drink often, or seldom. He seems to be capricious about it. Most herbivores want water every day, or every two or three days, but the giraffe seems not to crave it. If it is there, well and good. If not, he does without. Either way, he is content. He never makes much fuss about anything.

He is as celebrated for his peaceful, gentle disposition as for his height. His third distinction is his silence. It is generally supposed that the giraffe is completely voiceless, that he has no vocal cords, that he never, never, never makes a sound—that he *cannot.* A typical statement on the subject is that of Sir Samuel Baker, who was a most careful and experienced observer of giraffe habits. He says, "It is an interesting fact that giraffes are absolutely mute, and even in their death agonies never utter a sound."

Game Ranger Percival is one of the few men who has reported otherwise, and even he spent many years in Africa studying game before he made the discovery that giraffes were capable of sound. The first time he heard it, he mistrusted his ears. It was very faint, no more than a feeble bleat, but he thought he had caught it clearly. He wrote a number of letters, to find out whether any one else had shared his experience, and two men wrote in reply that they had. Since then, Martin Johnson, the photographer-explorer, and Dr. W. Reid Blair, Director of the New York Zoölogical Park, both say that the giraffe is not utterly voiceless. He has vocal cords, and while they are weak and atrophied, he can make himself heard at close range. It happens very rarely, but it is not impossible.

The World War and the giraffe—what possible connection could there be between the two? Would you ever suppose that the fate of any beast whose home was in the far-off African wilderness could in any way have been involved with the doings of kings and armies of Europe? Miles of land and the whole stretch of the

Mediterranean, and the sands of the Sahara separated the battlefront from the giraffe herds on the plains. Yet the War nearly put an end to them all.

It all happened because the giraffe is so tall!

Great Britain and Germany both had colonies in Africa. Troops had to be stationed on their border-lines, to spread their camps and make their marches and even fight wherever the two powers came into contact. One of the first steps an army in the field takes is to set up the telephone and the telegraph. Accordingly, tall poles and high-slung wires appeared on the plains.

Giraffes in flight have an astonishing skill in avoiding the branches of jutting trees. No matter how swiftly they gallop, they dodge and duck their long necks when they are going through tree country. Trees give them no trouble. They are used to things that grow. But they were entirely unfamiliar with such things as taut telegraph wires. As the herds thudded over the ground in full gallop, they tore headlong into the unseen wires. The cables burst with a crash. The poles fell under the shock. The giraffes dashed on, more frightened than ever.

The armies noticed with anxiety that their messages failed to come through! At first, each side thought the enemy had managed to destroy their wires, but they soon traced the damage to the giraffe herds. There was only one thing to be done, and every soldier had orders to shoot the beasts at sight. So many of the unfortunate offenders were killed, that the whole species was in danger of being wiped out. Only a few escaped.

Giraffes, however, have a strong chance for recovery from diminished numbers. In regions where they have been greatly reduced by disease, they have not altogether died out and in time they become as numerous as they were before the plague attacked them. At the close of the War, strict regulations were issued for their protection and their numbers have since multiplied. The "gentle giraffe" has survived.

4

THE OTTER

EVEN before the pioneers began their steady passage across our continent, there were occasional bands of men who broke through into the untamed country of the West. They were not in search of new homes. For the most part they led an unsettled, gypsy-like existence, traveling from year to year to and from the most favorable hunting grounds. These men were after furs, and their work took them toward the hidden trails marked by animal feet, but untrodden by man. A typical expedition of this kind carried about sixty men, though the actual hunting was done by but three or four. The others, Indians or half-breeds, looked after the horses, did the work of the camp, and frequently served as a small army of scouts and defenders against surprise attacks from hostile Indians. The married men often took their wives and children with them, for a fur hunt might easily take half a year, allowing for the slow pace of the pack train or the cart. There might be as many as twenty-five women and sixty-odd children in the party, and the outfit was like a small village in motion.

There were such troops of fur hunters in our Northwest only a hundred years ago, following a trade that had its origin far back in the earliest history of primitive man. It was the pang of hunger and the sting of cold that first led to the art of hunting. If food and covering could be obtained from the same animal, so much the better.

In the beginning each man hunted for himself and his family, but little by little things changed. The best farmers stayed in the fields to look after the crops; the skilled smith remained at his forge; and craftsmen and artists attended to the building and decorating of their dwellings. Only those who excelled in the hunt went in pursuit of the wild creatures who supplied them with meat and clothing, and they brought back their trophies, exchanging them for grain and materials with their tribesmen. In time, the custom of barter gave place to buying and selling with money. At first, any skin that was warm would do, but people came to want furs that were soft and pleasant to the touch as well, so that the hunters had to go farther and farther afield in search of animals with more beautiful coats.

One of the most highly prized of all furs since pre-Biblical days is that furnished by the otter, that slender, short-legged, alert and agile little animal of forest and stream. He is known on every continent except Australia, in both warm and cold climates. His narrow body, about a yard long from the wiry whiskers at his blunt muzzle to the tip of his tapering, but heavy tail, is built for speed. He is constantly in and out of the water, so that it is hard to say rightly whether he is a land or water dweller. He is amply protected against cold and wet by the layers of fat under his dense under-fur and by his longish, coarse outer hair, which acts as a sort of rain coat and enables him to stand prolonged immersion in the water. Its length and stiffness also aid in speedy swimming. The outer hairs are grayish at the base and brown at the tips. Many otters have patches of light gray or white at the throat, but most of them are solid brown in color.

In preparing otter skin for furs, these outer hairs are

removed, and the thick pile of the undercoat is dressed to a smooth finish. In this form it is highly prized. In the cold countries of the Far East it is still a favorite for men's coat collars and cuffs and, where long robes are still in fashion, they are often banded with otter in widths varying according to the social position of the wearer. It is also used on men's coats in Europe, though not as generally as it was in the days of cloaks and swords, and in America it is seen in the form of ladies' coats.

Although otters exist in so many countries, both tropic and northern, they have never been taken in large numbers. They are extremely swift in their movements and extremely sly, and the hunter often goes home empty-handed. In the long history of otter hunting all chroniclers agree that the "sly, goose-footed prowler," as Somerville, writing in 1735, called him, can be more baffling and tantalizing than any other forest creature. His short legs have their limitations, but walking is only one of his methods of locomotion. He runs, he jumps, he rolls, he even skids and slides along on his belly for a few yards now and then. His narrow, elastic body carries him in, over, under and through obstacles. He finds a loophole in a thick tangle of tree roots, among the rocks, under a slightly raised log, and in a flash he has wriggled away on a short cut. The otter is the star athlete, acrobat and trickster of the animal world.

In the winter, when the ground is piled high with snow, he has still another trick up his sleeve. His dark body shows clearly against the white snowdrifts, but he meets that disadvantage with strategy. With the hunter and the dogs at his heels, the wall of snow at his head, all escape seems cut off, but a short, quick leap lands him headfirst straight into the snow bank and his sinuous body cuts a tunnel as it moves through to the opposite end. The dogs stand sniffing and puzzled, while their quarry scuttles away to safety.

As a swimmer and diver, the otter excels. His clawed feet are webbed, which is an enormous help in his aquatic career. When he is swimming at a slow pace, he

OTTERS ON THE RIVER BANK

paddles without effort, dog-fashion, but when he wants to make speed, he throws his whole body into strenuous action. His powerful muscles swerve him forward in a curving motion, and his tail switches in smart, strong strokes, acting as propeller and rudder at the same time. When he is really in a hurry, he shoots through the water like an animated torpedo.

The otter takes to the water chiefly because he lives on

fish. In an emergency, or as a side dish, he may snatch a few mouthfuls of shellfish or a batch of frogs, but his real needs are to be satisfied only with good-sized, firm-bodied fish, freshly caught. He bothers very little with small fry, but goes after the heaviest and strongest creatures in the stream. He will dive after a sounding salmon, try his skill against a wily trout or grapple in a struggle with a fighting pike, though it may take several attempts before he lands his catch. Once he has caught his fish, he seems to recall the fact that he is, after all, a land animal much of the time, and he swims to the shore, holding his prey in his mouth, and climbs up the river bank to eat his meal. He eats quickly, and often leaves his fish unfinished, preferring the novelty of a fresh catch.

By and large, the otter has paid dearly for his cleverness as a fisherman, for he has been harried and hunted away from the choicest streams, in Great Britain and Ireland especially, to make room for men who came with angle and net. They refused to have their sport spoiled by the ravages of the otter, and removed the rival who depleted the stock of fish and interfered with their success. In many places otters have vanished by reason of the fishermen's grudge against them.

Early otter hunting was carried on exclusively for fur. Then followed the drive against them on the part of indignant fishermen. Lastly, they became the victims of the sporting gentry, particularly in England, who found the chase of so elusive an animal a challenge to their skill. For the pursuit of the otter, special breeds of hunting dogs were developed, whose chief characteristics were extreme acuteness of scent, strength to withstand the fatigue of a long hunt or swim and a shaggy coat of hair to keep off the chill of the water. Once the otter took to the river, their only chance was to swim after him to the point of exhausting him. King John, of Magna Carta fame, had a kennel of otterhounds, and many nobles and rich men of his day and the following centuries followed his example. Within the last fifty years otter hunting has died out, mainly because the animal has become so scarce.

Having caught their quarry, the otter hunters kept the skin as a trophy, but neither man nor dog had any taste for his meat. One huntsman reports having offered his hounds a chance at a carcass, but they turned away from it. His cats, however, took to it with relish. It has been described as having a rank, fishy flavor, and while it has been eaten by hunters when nothing better was to be had, it holds a low rank as food. Long ago, one of the monasteries of England listed otter meat as a flesh permitted on Fridays, presumably because of its resemblance in flavor to fish.

When the winter is severe enough to freeze the streams, there is no sign of the otter in the woods, except an occasional bubble under the ice. Diving after fish is impossible, but it is quite easy for him to slip into the water from underneath. In preparation for the coming cold, he chooses a hollow in the river bank below water level, and tunnels into it, pushing his way through on an upward slant to a distance of several feet. At the end of the passage, he establishes his den. The frozen roof of the ground keeps out the cold. The upward line of his tunnel keeps the water from washing into his quarters. When he is hungry, he need only slip down the chute, catch his fish and come back to the snug comfort of his home to dine. He swims at quite a depth, for he is a proficient under-water swimmer, and comes up now and then close to the icy crust for air, and it is then that the "otter bubbles" are seen. When the frost breaks and the stream is open again, he leaves his tunnel and takes up his roaming once more.

Over a range of twenty miles it may happen that only a single otter is tracked. They travel singly, as a rule, and make long journeys overland, following the course of the stream as all good fishermen do, looking for the best-stocked spots. The prowler is artful enough to take cross-country shortcuts when changing to better waters. Following the signs of his travel, which show in footprints, narrow grooves made by his sliding, and bits of fish and frog left over from his feeding, it appears that every otter has his own private territory. Somehow, he sticks to his own "beat," as though obeying an invisible sign marked NO TRESPASSING. The extent of an otter's exclusive range has been estimated at about twenty-five miles. No doubt their trails cross, but not for long. It is thought that the intruder is warned off by the scent of the "owner of the beat."

When a band of two or three otters wander about together, the company usually consists of a mother and her young, or of brothers and sisters who have not quite reached the age of independence. Two or three is the number in a litter, and the babies are kept secure in a nest made in the hollow of a rock or in the angle of a twisted tree stump, cushioned with heaps of grass. The mother nurses her babies, and keeps them at home until she decides they are old enough for schooling. Then she leads them to the river brink, scoops them onto her back and swims out to deep water. They seem to enjoy the ride, but suddenly they feel their mother's body slipping away from under them. They squeal in terror, but she disregards their cries, leaving them with apparent heartlessness to sink or swim. They swim, somehow, very nervously and with a great deal of frantic splashing. All the time the mother otter keeps a careful eye on her children, and at just the right moment, when they are too tired or too scared to keep afloat, she slips under them, supporting their little bodies, and instantly the crying stops. Every day she lengthens the lesson, and before long they swim more and more calmly and expertly. They next take up the art of diving, learning by copying their mother's form, and one fine day the youngsters find they know how to catch fish just as she does. For a time the group goes about together, on land as well as in the water, and when their strength and agility warrant complete self-reliance, the little band breaks up and each otter goes his own way fishing for himself and keeping out of danger as his quick wits dictate.

These are the ways of the land otter in the wild. There is another brief page in otter biography that shows him in his rôle as captive. Far away, in the north of China, on the bank of one of the small tributaries of the Yangtze River, there is an insignificant little settlement known as "Otter Village." It has never achieved the dignity of a dot on the map to mark its location, but for hundreds of generations of men it has had a special fame because there have always been a number of otters making part of the community. The Chinese of this hamlet live on rice and fish, and they employ their otters as assistants in going after the catch. The animals are trained from infancy, not only to do their share of the work, but to refrain from eating the fish until the master throws it to them. On fishing days, man, otter and net go out together in a one-oared sampan to a quiet spot in the river. As the fisherman casts his net, the otter plunges overboard in a clean dive and swims about in the deep water, routing the fish out of their hiding places and scaring them toward the sunken net. If the otter comes up promptly, the man knows that fish are scarce and that he must paddle away to try his luck elsewhere. Down goes the net again, in plops the otter, often coming up in the net, together with the fish.

How old the practice of otter fishing is, no one knows. It was first written about by Chang Tsu, an author of the Tang dynasty, who describes a second method, used by the less ambitious fishermen. Their otters dive in, unleashed, and bring up the fish, one by one, in their mouths, swimming to the shore and dropping the catch at the master's feet. In Otter Village and in a few remote spots on the Ganges and here and there on some of the Malayan streams, there are men who still fill their baskets in the slow, patient, and somewhat lazy way of their ancestors.

When the otters are not on active duty, they are kept tied to a post of the fisherman's hut. They are pleasant, good-natured animals, with no viciousness about them, and make gay playmates for the village children.

Even in the wild state, otters show a love of play that is almost human. Most animals have some sort of sport, especially when they are young. They tease each other, skip about, roll on the ground, try a little wrestling or boxing, or play tag and something like stump-the-leader. The otter, when he wants a supremely good time, goes coasting. He picks out a hill with a steep incline dipping down to the river. In the summer, the earth is damp with the forest moisture. In the winter, it is smooth with snow. Flat on their bellies, the otters go whooshing down the slide, shooting into the water with a splash. Up they scuttle again to the top of the hill for another thrilling dash. After a few speedy trips, the snow on the hillside turns to glassy ice, catching the water from the otters' wet bodies, and the pace of tobogganing grows faster and faster. The shining, brown animals keep at the sport untiringly, as was noted by a hidden onlooker who saw a pair of otters take the plunge downhill thirty times, stopping only when they caught the clue of his presence and slithered away.

Alaska, more than any other part of the world, has reason to recognize the existence of the little land otter's big relative, who carries a family resemblance, but is quite distinct from him in appearance and behavior. His home is the sea, but he is not strictly an ocean-going creature, for he does not frequent the far depths. The sea otter's beat is confined to the north Pacific, now in a small area, though not so long ago he splashed about among the kelp beds edging the shores of the islands of the Bering Sea and skirted the rocky ledges of the coast on the American side all the way from Alaska down to Lower California—but this was before Peter the Great, of Russia, had his enthusiastic talk with his ablest navigator, Vitus Bering, on the subject of the nature of the land that hemmed in the ocean on the wall opposite Kamchatka.

Peter the Great had ideas for the enlargement of his empire, and, being himself a lover of ships and interested in navigation, he formed a plan with Bering for an expedition across the Pacific. Peter died before the plan was carried out, but Catherine the Great went ahead with it, and in 1741 Bering took his fleet of two ships on the voyage of discovery. One of the ships was forced to turn back, but Bering sailed on and struggled to a landing off the Alaskan coast. His supplies ran short, and between hunger and the biting cold, the explorers had a hard time of it, until the desperate crew took to hunting the fat, thickly furred animals that came within their reach, approaching the shore in search of mussels, cray- and shell-fish lodged in the rocks at the foot of the waves. Unlike the "true," or land, otter, the sea otter does not pursue the strong, finned fish, but lives on crustaceans mainly. He is a foot or two longer than the otter of the woods, and more heavily built. His hide covers him generously, wrinkling in thick folds around the neck especially.

The sea otter not only saved the lives of Bering's men, but tied a ribbon of contact between America's coast and the countries on the opposite line of the Pacific. The sailors stilled their hunger on his meat, and wrapped themselves in his fur, cutting it into crude coats and blankets, hanging the pelts across the wind-battered walls of their camp and stuffing them into the cracks left by their rude carpentering. They little dreamed that the animal that served to fill their starkest needs would one day be known as the most valuable fur bearer in the world.

When the Bering ship returned across the Pacific, landing on the coast of China, the sailors stamped ashore, themselves and their dunnage bundled in sea otter pelts. Rumpled and dirty as the furs were, they caught the attention of the Chinese merchants, for they were of a thickness of pile, a softness and beauty unequaled by any other peltry. The sea otter is deep brown in color, with tones of blue and black enriching it and a sprinkling of little white hairs throughout. It is glossy and luxuriously dense. Its most unusual feature is that the under fur and the outer guard hair are very nearly the same in texture. Bering's crew sold their coats and blankets and wall coverings at prices that made them dizzy. In no time the news spread westward to Russia and Europe that there was wealth swimming about in the north Pacific, and there was a frantic rush on the part of adventurers and treasure seekers toward the Kamchatka coast where they took ship for the new-found hunting waters.

The men who embarked on the fur hunt were in the main rough and hard, equal to the brutality and strain of life in the open, cold sea, but they were unskilled in the hunt. But along the shores of the Aleutian Islands, where the sea otters were thickest, they found the natives, expert with the spear and quick with the skinning knife, willing to work for them. Rocking in their little, low, skin-covered boats the Aleuts risked their lives, toiling under the bullying and driving of their masters, who stood safe on their high decks shouting orders and curses down on them, urging them to more and more active spear-plunging.

Back and forth between the walls of the Pacific the fur traders sailed, growing rich on the velvety cargo. Some of them, tiring of the hard life, took their earnings and settled in Alaska. As they moved inland, they found even greater riches in the inexhaustible salmon waters, in the timber forests, in the mines and in the teeming wild life that could supply furs for the trapping. Alaska woke to life. She remained Russia's property until 1867, when the United States purchased the territory, paying over seven million dollars for it, and though it seemed a large sum at the time, Alaska's unbounded resources have since repaid the investment immeasurably. While the settling of the peninsula drew more and more men to our Northwest, the assiduous hunt of the sea otter off the coast went on, declining only when the work became harder and the rewards fewer as the animals became scarcer.

To-day there is no such thing as sea otter hunting on

our side of the Pacific. It is only a memory, not an industry. The fur of the sea otter is no longer a luxury; it is a rarity. On the Kamchatka side three or four dozen pelts may be taken in a year, with luck. On our coast, the hunt is not only forbidden, but next to impossible. Here and there among the islands of western Alaska, a few pair of sea otter still exist, rare reminders of the thousands that once sported on the edge of the ocean. The United States Senate has received the report of their whereabouts from its Committee on Conservation of Wild Life Resources, with an urgent recommendation for the establishment of a "water preserve" off the Pribilof Islands, where the animals once flourished in such numbers. It is proposed that the few surviving sea otters be taken by net and transplanted to this favorable locality.

Another solution to the sea otter scarcity would be the acceptance of Japan's offer to sell the United States a number of the animals from the rookeries of her coast line, at a price of two thousand dollars each, or on the basis of exchange for some of our Alaskan fur-bearing animals which Japan would like to introduce into her own territory. The Committee advises using our own sea otters in preference to the Japanese, even though they have to be netted and transported, as the nucleus of the colony. If it is properly stocked and protected, they will undoubtedly multiply sufficiently to forestall the extinction that threatens them.

When the sea otter hunt was at its height, it was no unusual matter for a single expedition to take from five to ten thousand skins a year in one locality. The hunters could at first keep close to the rocks edging the ocean rim and the reefs a short way out, but after a time the animals changed their resting places from the shore line to the floating kelp beds beyond in an effort to escape their pursuers. They have much of the shyness of their land otter kin, but show signs of curiosity that makes them rear their heads from the water for a long stare before they resort to a disappearing dive. Their playfulness takes the form of lying on their backs in the water and tossing pieces of kelp into the air, according to the stories of men who have watched them. There ought to be room in the Pacific for so harmless a sport as this!

5

THE KANGAROO

ONE hundred and forty-six years ago, facing unknown conditions, a small band of people landed at Sydney Cove. These were the first regular settlers of the continent of Australia.

Dazed by their long sea journey from England and dreading what they might have to face in this new land, the newcomers must have felt rather like Alice through the Looking Glass. For in this new land of their exile everything was as different as it could be from the country they had left behind. Here June, July and August are winter; here instead of the soft forests of England which lose their leaves in winter, they saw the high, gray-green gum trees green all the year round, and ferns fifty feet high. Instead of the quiet, delicate harebell and primrose, there flamed the brilliant yellow and red Australian flowers. Here there were no horses, no pigs, cows or even cats. Instead, these people were to make the acquaintance of Wonderland animals—strange pouched creatures who ran away from them in great leaps; animals who laid eggs and had bills like ducks, and a fierce, unfriendly native dog.

But foremost, the kangaroo, that ungainly gray creature who now shares with the emu the place of honor on the coat of arms of its country, the Commonwealth of Australia.

The kangaroo they saw was probably the Great Gray,

for the Great Red does not occur within several miles of the sea, and the early settlers only knew that strip of the Australian coast where a landing had been possible, a narrow strip, hemmed in, in back, by mountains.

It is hard to tell the Gray from the Red except by locality, particularly as the females of both are the same blueish-gray color, and as you may meet the Gray inland, even although it is a coastal form. If you can get near enough to see, you find that the upper lip of the Gray is covered with hair, whereas the muzzle of the Red is naked. In the males of the Red, the thick, plushy fur is reddish brown above; the legs and undersurfaces are white.

This strange creature is Australia's most characteristic animal, and once meant food and clothing to the natives—and a kind of religion as well. Many of the Australian tribes still believe that in the very distant past, in a sort of dream world, there lived a people from whom they are descended. These people, in the legends of the natives, are so mixed up with animals that the natives regard them as part human and part animal. Members of the ancient group whose totem was the kangaroo are spoken of either as the "kangaroo man" or the "man kangaroo." And even to-day in some tribes a kangaroo man may only marry a kangaroo girl—that is, a girl from a tribe whose totem is also the kangaroo. The ceremonies bound up with the kangaroo are many and in these the appearance and motions of the kangaroo play a prominent part.

So unlike any other animal in the world is this large creature who leaps about on its hind legs, that the early voyagers to Australia were quite at a loss to describe it to their friends at home. It looked like nothing but itself. They usually made the best of a poor job by saying that it seemed to be a huge rat of cow-like proportions, with a head like a greyhound, and a gait like a rabbit! The earlier voyagers, however, who went out from the great trading companies, did not linger to look at the kangaroo. The natural barriers of the Australian coast were too much for them. They were content to describe the land as rocky and impossible, its native inhabitants as the "miserablest people in the world," its coast line as without possible harbors, and to sail away, leaving behind them, behind those barrier mountains, the fertile valleys of an island continent as large as the United States.

Strangely enough, it was the American Revolution that seems to have started up regular settling in Australia. Up to the time that America freed herself of British rule, England had been in the habit of disposing of her undesirable citizens by dumping them on her American colonies. When this refuge for both convict and political undesirables was cut off, she was obliged to seek some other place. Thus the early settlement of Australia was begun by such refugees and their escorts. One famous voyager, Captain Cook, had previously found that there was a fertile landing place on the island continent, and had reported it to his government, and with this knowledge, the first colony was dispatched to the unknown country. They stayed within the strip of coast backed by the Blue Mountains, and not until much later were the barriers of these mountains overcome by men of a different status and a spirit of enterprise undimmed by jail—governors and their staff, and a few settlers who had obtained land grants from England on condition that they invest capital in the colony. These men tried the usual experiments of colonists in a new country—cultivation and stock raising—and soon found that the small settled area was too small for grazing purposes. Exploration of what lay on the other side of the mountains carried with it the usual tragedies, and at least one unsolved mystery in the death of a Prussian scientist who set out to cross Australia from east to west, but never returned and of whom no authentic trace has ever been found.

KANGAROO WITH SMALL KANGAROO IN POUCH

It was early discovered that beyond the mountains lay a great variety of land, both fertile and arid, and that to get to some of the fertile land, enormous tracts of desert must be crossed. The problem of transportation was grave.

None of the animals native to Australia are domestically useful. The kangaroo, although so strong and muscular, cannot serve as a horse or a mule, and the platypus, although it lays eggs, is no good as a substitute hen. All the useful animals in Australia had to be imported.

Along with the cattle and horses, for the special purpose of desert transportation, camels were imported from India, with their Afghan camel drivers. They are still in use to-day, for the Australian desert is vast and waterless.

Before the camels came, the land explorers had faced starvation in their efforts to master the wilderness. And often when their food ran out, that useless animal, the kangaroo, at last served a human need. As its meat had fed the unfastidious palate of the native, so now it fed the starving white men. One after the other of the early adventurers record running out of meat, and note in their diaries that "the men have been living on kangaroo meat for several days."

However, in spite of hardships, a new world gradually opened—a world of forests, mountains, deserts and fertile lands. And among the sugar cane on the great alluvial plains; in the wet dark jungle of huge trees from whose branches twine lianas and orchids; up among the steep rocks of the mountains, jump, burrow and climb members of the pouched tribe, carrying their young with them in their pockets.

These are Australia's marsupials—a word taken from the Greek word meaning pocket. They are practically Australia's alone, for the opossum of America is the only important representative outside.

The kangaroos are the best-known and the largest. They reach a height of seven or eight feet. From these, the family ranges down in size, through the wallabies, which are merely small-sized kangaroos, to the very tiny musk kangaroo whose head is smaller than a rabbit's. The family even has an underground member, the pouched mole. He is a rather pathetic little object, blind and covered by long, silvery hair. If you pick him up he will keep right on trying to burrow through your hand, his small blind head pushing and his small front paws digging, although there is nothing to dig.

Before the white man came to Australia and brought with him the animals of his own country, this land swarmed with pouched creatures large and small, leaping, running, hiding in the caves of the mountains, burrowing into sandy soil, or balancing high up among the trees. Whatever the place may lack in variety of the kind of animal it owns, it amply makes up in the uniqueness of the native stock. Every sort of disposition is represented in the kangaroo tribe, from the meek timid kangaroo to the friendly little koala, and the fierce and unfriendly Tasmanian wolf of the mountains. All these animals have native names which make for much confusion—the koala, for instance, is sometimes known as the native bear; the cuscus of the islands north of Australia is called the opossum; there are native cats, rats, wolves and even rabbits—none of which is really what it is called.

Their food varies as much as their dwellings and dispositions. The Tasmanian wolf and the Tasmanian Devil are flesh-eaters, as are the native cats and rats, and destructive to sheep and fowls; other marsupials eat grass and leaves, like the kangaroos and wallabies; others eat roots, like the wombats. The phalangers eat fruit; the honey mouse eats honey and insects, and the bandicoots and mole eat insects. One relative of the kangaroo, a tree dweller, lives almost entirely on honey. This is the honey mouse. Its method of getting the food is to stick its long, rough tongue into flowers or to curl it around them. This tiny, rat-like little animal, with its long prehensile tail and triangular face makes a small grass nest in the trees where it spends its life. When the Ti Tree is in bloom, the honey mouse lives there; when the bottle brush is in season, the honey mouse goes down to the wet flats where the spiky bush grows. As some variety of the bottle brush is in bloom nearly all the year round, the honey mouse is in no danger of starving in spite of its restricted diet. It is thoroughly at home right

THE TASMANIAN DEVIL

side up or upside down in the trees, and when it sleeps it rolls itself into a compact ball and you cannot easily make out which end is which, or whether there are any ends at all! There is another small animal that travelers occasionally encounter in Australia, also a kangaroo relative, who has the same long tongue and uses it to eat insects with. It is very shy and people who have seen it say it usually tries to run away and hide in the hollow of a dead tree, but that before it does this, it will stand up on its hind legs and look around, as if to see whether the retreat is really necessary or not. This is the animal confusingly known as the Banded or Striped Ant-Eater. It is not an ant-eater, but simply a marsupial who eats insects. In spite of the fact that it has more teeth—fifty-four—than any of its relatives, it is a harmless animal, and if captured does not use these weapons in its own defense. Its cousin, the wallaroo, although not possessed of such a fine assortment, uses its teeth in fighting.

As the chamois climbs without hesitation among the peaks of the European mountains, so the rock wallaby skips over the mountains of Australia, its rough toes giving it a sure hold even on the most slippery surfaces. This pouched animal follows self-made trails which are practically invisible to man but which it finds and uses swiftly and surely.

Another very sure-footed kangaroo relative is the tree kangaroo who, among other climbing accomplishments, can make downward leaps of sixty feet or so, landing without injury!

The matter of sure-footedness, speed and defense is very important for a wild animal, and becomes increasingly so as man encroaches on its territory, bringing with him firearms and the dog. All the accounts of the early land travelers in Australia speak of the natural enmity that exists between the kangaroo and the dog, both the dingo, or native dog, and the imported, domestic variety. The kangaroo, who on the whole is a very peaceful animal, has a unique and very telling method of defense. It is a natural boxer! Male kangaroos often stand up to each other and box for half an hour or so, and the animal can be trained to wear gloves and box with human beings! Much as other animals, like bear cubs or kittens, cuff each other in play, so the kangaroo boxes in play, but with his weak-looking forefeet he can administer a punch that may not feel at all playful to its recipient. In real fighting, the kangaroo's technique is quite different from his play boxing. In a fight, if possible, he

THE HONEY MOUSE

backs up against a rock or a tree, standing raised on the toes of his long hind feet and balanced on his tail as on a third leg. Normally, he rests on his hind legs as if we were to rest on our whole forearm from the elbow down, but in fighting he rises on his toes, lifting what seems like a back elbow, but really is his heel, from the ground. Balancing on his stout tail, he then grasps his enemy with his forelegs and holds him tightly, while, bringing a hind leg forward, he swiftly inflicts a ripping horrible wound on his adversary with the stout sharp nail of the fourth toe.

The attacked kangaroo is very apt to flee to water, if there is even a pool near by. Standing there, he will go through this performance. This has given rise to the story that the animal deliberately carries his enemy to water, and as deliberately, drowns him. The truth is that if he can get to water, he will; if his enemy is near the edge, he may be dragged in by the kangaroo. If the enemy pursues the kangaroo into the water, he is likely to be grasped in the usual way, and very likely will get such a kicking and clawing that he will not emerge again, either because he has been pushed under the water by the kicks or because he is too wounded and exhausted to move. But not because the kangaroo thought out the situation beforehand and carried out a well-formed plan to drown his enemy.

This fierce fighter is not a pugnacious animal by nature. He is timid and occasionally has been known to die of fright! He is seldom seen near settlements but he does sometimes appear in a small group of his kind in the vicinity of outback farms, and unfortunately he has a habit of invading crops. He can be tamed easily and will follow his master or mistress around like a dog, or play peacefully in a back yard. Providing there is plenty of room, he does not seem to have any objection to captivity.

Enormous and ungainly as the large kangaroos are, when they are born these same creatures are less than one inch long. They are, at that important time, most unattractive. They are tiny, blind, and entirely without the dense grayish fur that covers their parents. But weak and blind as they are, they manage to migrate up through the mother kangaroo's thick pelt into her pouch, where they settle down to protection and nourishment until they are clothed in fur and strong enough to look after themselves. Usually only one kangaroo is born at a time and the pouch of the typical kangaroo is adapted to hold only one.

All the marsupials have these pouches, except the

THE BANDED ANT-EATER

banded ant-eater. The rabbit bandicoot and the marsupial mole, both of which, unlike the kangaroo, travel on all fours, have pouches that open to the rear. This prevents sticks and other obstructions from entering the pocket when the female is on the run. In some marsupials who have larger families than the kangaroo proper, the pouch can accommodate many more than one child.

There is a story that if pursued with a young one in her pouch, the female kangaroo, finding her pace slowed down by the extra weight in front, lifts the child out of the pouch with her forepaws and casts it to one side in the bushes. She is then supposed, when danger is past, to return to the spot where the sudden parting took place and gather up the child she has temporarily checked there.

The kangaroo's forepaws are not adapted for lifting the young kangaroo out of the pouch or anywhere else. They do not even help the inch-long baby on its first weary journey into the pouch. If the mother kangaroo bends and jumps sufficiently actively, perhaps the young wriggle or fall out of the pouch. And possibly if she passed that way again, the elder kangaroo would collect her young, or instinct might lead her back to it as a cat finds a lost kitten. But there seems little authority for the assumption that she actually reasons out the danger, lifts out her child, and then returns to pick it up.

The baby kangaroo which nurses while in the pouch, is so firmly attached to the mother when feeding during the early part of its babyhood, that many people believe it was born in the pouch, which is not the case.

It takes several months for fur to develop and for the little kangaroo to emerge from the pouch. But even before it makes this first voyage to the outer world, it has begun to nibble grass and leaves. When the mother is feeding and bends low enough for her child to reach the grass, the small, pointed head will emerge from the pouch and nibble a little too. But even after the young kangaroo has learned to get out of the pouch and go around alone, if frightened by anything, he will rush headlong back to the pouch. Actually headlong, for he dives in head-first, hind legs and tail waving in the air until he has managed to wriggle around and get right side up again. The mother kangaroo keeps her pouch clean with her tongue and her front paws, but the kangaroo infant is not always kept as clean as the pouch. It is in fact a rather dirty little child, although otherwise it is well cared for by the gentle mother animal.

The male kangaroo appears to be completely uninterested in his child. His chief interests seem to be food and rest. All the kangaroos rest quite a bit during the daytime, lying placidly on their sides in the shade.

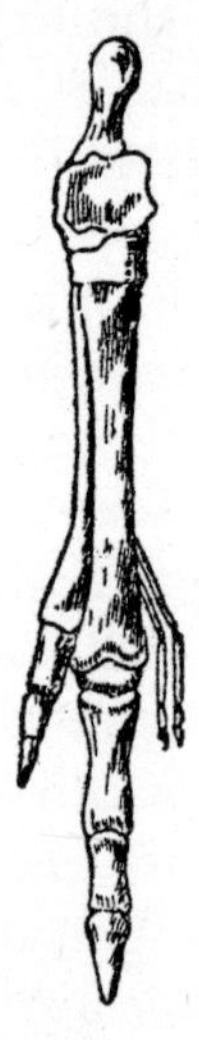

SKELETON OF KANGAROO'S HIND FOOT

When a mob of them are frightened, the females and the young ones get away first. The old males go more slowly. A kangaroo herd in flight is very beautiful. Their large, long, gray bodies almost fly over the ground.

Among the herds there are usually some "boomers" or "old men"—old males that have grown enormous and heavy. We grow to a certain age, maybe sixteen or eighteen, then we do not get any taller. But the kangaroo, although slow about it, keeps right on growing. Some of these old males are immensely heavy and tall. Occasionally, one of them will suddenly bound down the street of a small village, covering the ground too rapidly for the astonished onlookers to do anything but stand and stare while his gray body leaps past them out of the village and out of sight.

Moving around slowly or grazing, the animal uses all four legs in an awkward ramble, but when it wishes to make speed, it uses only the hind legs, holding the short forelegs close to the sides. It progresses then in a series of powerful and not ungraceful leaps. Balanced by the tail, which is very stout and strong near the body and keeps the animal from toppling over backward in the jump, it makes smooth bounds, four or five feet in length. These leaps can be as long as nine yards, and the kangaroo has been seen to take a nine-foot wall, probably under the impetus of extreme fright.

The typical adult is built for its jumping method of progression. It has very small, short forelegs and extremely long and powerful hind ones. Its haunches are large and muscular and its feet very long.

But when this same kangaroo is born, the legs are just reversed in proportions. The forelegs are much the longer and stronger, with large claws, and the hind legs are very short and very weak. As the animal grows, the forelegs stay short, while the hind ones grow longer and very strong, with long narrow and homely feet, armed with heavy claws, the fourth of which is longest and sharpest.

There has been considerable depletion in the ranks of the kangaroo because of his hide which makes excellent shoe leather. He is also killed because he ruins crops, and he may go hungry because the imported rabbit nibbles away his food. But the kangaroo's fiercest enemy remains the dingo. In times past the dingo was without doubt the worst enemy the kangaroo had, but now it is a question between dingo and man. Unfortunately for its own peace, the kangaroo is a vegetarian, and like most vegetarian animals is very fond of gar-

"JOEY"

den produce. It is a highly unpleasant surprise to a hard-working farmer to drive past his wheat field one fine day and suddenly encounter the mild, dark brown eyes of a kangaroo whose long, gray head sticks up innocently above the high wheat he has been engaged in depleting. The farmer's first impulse is to shoot the thief, and in spite of severe government restrictions upon the destruction of this unique animal, this impulse is often followed and a great many kangaroos lose their lives because of their taste for home-grown crops.

No one can think of the kangaroo in Australia without remembering its near relative, a small marsupial known all over Australia as "Joey." This is the koala, a little animal who lives chiefly in the trees which it seems to climb with some effort. It must have been the original model for the Teddy Bear. Its ears are large, furry and round, standing out widely from the head. Its nose is funny and broad and black, and on either side of it the round eyes gaze out with a dreamy stare. It has no tail and its furry little body is fat. If it traveled well, which it does not, we would certainly know it as a popular household pet. In Australia it is very popular, and so irresistible that you can even forgive the digs you get from its claws as it climbs you as if you were its native haunt—the eucalyptus tree. The koala's front toes are arranged so that it can grasp the branches it spends most of its life among. Its soft, dense fur always smells strongly of the pungent leaf of its home tree. But in spite of the power of eucalyptus oil to ease sore throats, this animal who lives on it has a strikingly harsh and unpleasant voice. When you first hear it, you are startled that such a loud, raucous crying could emerge from such a little lovable animal—it is undoubtedly one of the loudest, hoarsest voices in Australia!

The joey child lives for a while in its mother's pouch, but very soon the mother begins to carry joey around on her back, making her slow progress up the gray bole of the tree, where she will sit and stare down at you dreamily, the baby fast asleep on her back. Her responsibility for her child seems to end there. If you take the joey off her back, she goes right on dreamily chewing gum leaves as if nothing had happened.

If you go into a fur shop over here and ask to see a piece of wombat fur, you may be pretty sure that you are really looking at koala. These two very different animals have been much confused by furriers and people in the fur trade, and the "wombat" of that trade is almost always the koala. You can tell that it is because the koala's guard hairs—the long glossy hairs—are pale gray with faint white mottlings, and the fur fibers, or softer, woolly and very thick hair surrounding the guard hairs, are very dark grayish-brown. On a whole skin, near the lower part of the rump, you will see a small patch of white hair, shorter than the other hair. The real wombat is larger than the koala; its fur is very coarse, and although it ranges in color from yellow to black, the skins, which are bought for leather, not for fur, are apt to be dark brown. In Australia, dozens of men have been sent to jail because they have been caught in possession of koala skins. But in spite of all the efforts to protect it, this little animal is becoming more and more scarce. The little sluggish creature is much easier to capture than its large relative, the kangaroo. Kangaroo-shooting requires skilled marksmanship, and the kangaroo population of Australia is being not so much actually depleted as it is pushed back into the wilds by increasing settlement. In spite of restrictions, it has become so unpopular with the farmers that in one way or another it has almost entirely disappeared from many districts and today there are living in the towns of Australia hundreds of people who have probably never seen a kangaroo in the wild, although not so very many miles away from them in the bush there are still thousands of this unique animal.

6

THE PORCUPINE

ALTHOUGH the American Indians in their prime were a superb race, strong and inured to the rigors of life in the open, they had their share of aches and pains and illness. There were war-wounds and snake-bites, hurts sustained in the accidents of the hunt and all the fevers and diseases that bring suffering to men. For their healing, they turned to the medicine man. It was his task to cure the sick, but he was by no means the sort of doctor we know. He was expected to care for the health of his tribe, but he had also to use his arts to call down rain in times of drought, to invoke success in times of war, to placate the winds and storms and to ward off disaster generally.

The medicine man's learning was a medley of natural history, including a sort of practical botany which told him the uses of plants, a knowledge of the ways of birds, animals and insects and shrewd observations of the weather changes. In treating the sick, he did what he could with the various herbs and plant juices and many of the minerals, but for the rest he relied on magic. He had no idea of the cause or course of an illness, and was unable to tell whether the trouble lay in the lungs, the heart, the stomach or any other part of the body. The usual explanation of sickness was that somehow, perhaps when the person was asleep, a foreign presence, usually an animal, had entered his body and taken possession of it. The medicine man mixed potions and powders, but his main concern was to remove the disturbing animal or spirit, frightening it away by chants and ceremonies and working with his hands to "draw out" the evil, believing that if he banished the stranger from his patient's interior, the sickness would cease.

The Indians had no concrete names for the various diseases. When a man lost control of his muscles and tossed and thrashed about, they said, "He is behaving like a bear." This was the common description of a fit or convulsion. Of an insane man they said, "He is acting like an otter," referring to the quick, nervous movements of his body. When the patient gave signs of acute pain, they said, "There is a porcupine inside him." This seemed a most convincing explanation, for every Indian knew from first-hand experience what poignant agony a porcupine can inflict. Though the animal is seldom active and about in the daytime, the Indians were familiar with his hideouts, and when they went to hunt him, as they did for his meat and quills, they peered along the trunks and branches of the young trees on the lookout for a beady-eyed, small-muzzled creature about a yard long, distinguishable from all other animals by the thicket of quills rising on his hairy person.

The quills and hair vary from yellow to all shades of brown and from gray to black, and at a distance the porcupine looks like a thickened blur on the branch. Ordinarily the quills slant backward among the wiry hairs, but when he is frightened or ready to fight, they stand nearly upright, like a battery of tiny lances, making him look much larger than his twenty or thirty pounds would warrant.

As they stepped sure-footedly through the woods the Indians kept an open eye for stray quills lying about, for they knew that in his rambles a porcupine is very apt to drop a quill or two as he brushes along. They are loosely set and come out easily. Each one is tipped with a minute, jagged barb, piercingly sharp and so rough-edged, although the roughness is invisible to the eye, that it gives the quill a fast hold wherever it is lodged. It is useless to try to shake it out, and pulling it out is a form of slow torture.

The Indians made the most of the porcupine, finding his meat a delicacy when roasted and his quills a medium for one of their arts. They peeled the quill into thin, flexible strips, which they dipped into brilliant dyes and then threaded into their weavings in gay combinations of color and design. The pointed barb took the place of a needle and made it easy to run the quill as they wished. When the white settlers came among the Indians, quill work went out of style. The settlers found they could win the goodwill of the braves with presents of beads, and the bright colors and tinkle of the new ornaments, together with the fact that the beads were so much easier to apply than the quills, put an end to the appearance of the porcupine's weapons in decoration.

The porcupine leads a self-centered life, sticking to his own stubborn ways, usually traveling alone, and he is never the aggressor in a fight. Nevertheless, he is the least popular of animals. He is known from the Rockies to the Himalayas, and is familiar to people of both hot and cold countries, and is everywhere detested. The Eskimos who work with our fur traders in the Northwest know him and the Africans and men of the South American forests are acquainted with him. The African porcupine is startlingly unlike his relatives in America, for his quills are more like a fountain than a thicket. They run from ten to twenty inches and more in length and are banded alternately in white and black. The Africans put them to use, not as ornament, but in another of the arts, transforming them into a musical instrument to accompany their dancing. They spread the quills in two layers over a plank of wood, and strap them loosely in place with leather thongs, and between the two walls of quill they slip twenty or thirty "cat's-eye" beans, finally pulling the thongs tight. The combination results in a spirited rattle, as the beans shake rhythmically and resonantly against the hollow quills. The African porcupine spends most of his time on the ground, and feeds and roams during the night, out of sight while the sun is up except when he is routed out by hunters and their dogs.

The grudge against the quill-pig is strongest in India, where from the edge of the Himalayas to the southern tip of the peninsula he adds to human unhappiness, particularly among the English residents. When an Englishman goes to a foreign land to live, he sets to work at once to make his new home as much like a bit of his native land as possible, especially in the matter of his garden. The English countryside is famous for the beauty of its hedges and meadows. The humblest village has its cottages set with flower beds, and in the dingy mining districts and city slums there is hardly a windowsill without its flower pots, flaunting their gay color in the face of the soot and the fog. It is as though some drop of Druid blood still lingered in every Englishman's veins, compelling him to the worship of things green and growing. Away from home, he wants something of the same beauty about him. Whether he has a handsome house in town or a rough shack in the wilderness, he turns to his garden for a reminder of home and for solace against the loneliness of exile. In jungle country he clears a space and tames the chaos of the wild vegetation into order. Where the soil is arid and reluctant, he enriches it and persuades it to bloom. Not even the desert discourages him.

In India, where many colonials have their permanent homes, the houses stand in compounds, lovely with a profusion of native flowers and imported English perennials competing in beauty. The tree trunks are clustered with orchids. A less conspicuous corner holds the vegetable patch, tended as zealously as Robinson Crusoe's corn and barley sprouts, which grew from the kernels he accidentally spilled from his sack. Both for the sake of health and for enjoyment the Indian colonial toils devotedly over his home-grown vegetables. It is not easy to make them prosper in India, where the seasons vary from months of drowning rain to months of parching drought, and a successful truck patch is a triumph.

One fine morning the gardener awakes, and pays a visit to his beloved vegetables, only to find that the toil of weeks is undone. The patch is a ruin. It is as though a blight had come upon the young sprouts in the night, a living blight, armed with greedy, merciless teeth. The stalks are stripped of their buds, chewed and bitten through and through. There is no doubt as to who is guilty of the wreckage. It is the work of that universal instrument of destruction, that pest, the porcupine.

His teeth are the badge of his kind. He is one of the rodents, or gnawing animals, and has their characteristic front teeth, curved, long and powerful, one pair in each front jaw. They are about an eighth of an inch wide, exposed to a length of half an inch, and shaped at the tip like the edge of a chisel. They are brilliant orange in color. Except for them, his front jaws are empty, and a gap runs from them to the little trio of grinders on each side. It is these orange tools of his and their destructive work that have made the porcupine so hated. The bark of young trees is his favorite food, and he is almost equally partial to the firmer vegetables, when he can get at them. Owners of timber and of plantations and gar-

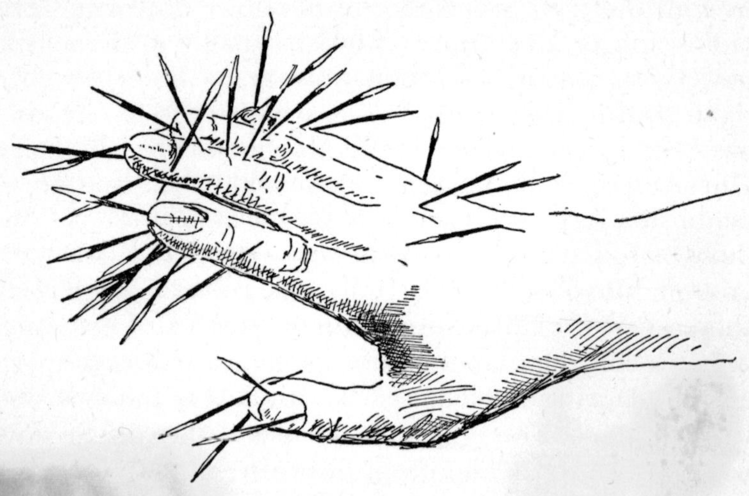

AN ENCOUNTER WITH A PORCUPINE'S QUILLS
(*drawn from a photograph*)

den lovers all over the country suffer loss and heartache because of him. The government has outlawed the porcupine in more than one district, setting a reward of two rupees for every one delivered into its hands. A great many are captured and destroyed every year, but they seem to grow more and more numerous. Every porcupine seems to have nine lives, or, at least, enough descendants to keep the population on the increase. In spite of the long war waged against the prickly pest, he flourishes. The use of traps and poison in private grounds is risky because of the danger to dogs and other pets. Capture by other methods is difficult, as the Indian porcupine is peculiarly wary and swift.

His mode of housekeeping is contrary to porcupine custom elsewhere. In most countries the quill-pig has no housekeeping at all. The South American porcupine lives in the trees and rarely comes down to earth. His tail is longer than that of others, and has prehensile powers, so that he can curl it around a branch with a tight, secure grip. He is the smallest of the family, and has shorter and fewer quills among his bristly hairs. Our American porcupines seldom take the trouble to establish a residence for any length of time, either. But those of India have their homes the year round in underground burrows inside a mudbank or hill, so cleverly chosen that the doorways are unnoticeable. They often live in a cluster of families, occupying neighboring apartments. During most of the day they stay indoors, and if one comes out for an occasional nap in the sun, he stays near enough to the mouth of his burrow for a quick getaway. Any native who hopes to get his two rupees by outwitting a porcupine is a fool for his pains. No quill-pig will venture out as long as there is any living thing near. He has a sharp sense of smell, and inside the clay walls of his dungeon he sniffs any foreign presence and cagily stays under cover until his nose assures him the coast is clear.

In high-grass country he often comes out during the day, shooting out of his burrow and into the shelter of the grass with galvanic speed. His legs are short, but he uses them to the best advantage, and on his journeys he works up a speed of ten miles an hour, which is a lickety-split pace for any forest creature. No one has been able to explain the extraordinary hurry that seems to possess the Indian porcupine, for his fellows elsewhere are slow travelers. His chief enemy is the leopard, but even where there are none about, the porcupine scurries through the forest as if demon-pursued.

In recent years a new sport has developed, which is more fascinating than any other connected with wild animals, that of photographing them in their natural habitat and in unconscious poses. Stalking and hunting a wild thing is exciting work, but catching him with a camera is doubly so. It is more dangerous than hunting, because the photographer is so intent on getting a picture that he has no time or attention to spare for self-protection. He is always accompanied by a gun-bearer, who keeps watch while the camera-man works. The best animal photography is done in the dead of night, when the animal world is most alive. Sometimes the creatures take their own portraits by stepping on a hidden wire laid so as to connect with a flashlight and strung close to the favorite bait of the individual. This leaves too much to chance, however, and the real enthusiast sits up night after night, impervious to the chill of the damp jungle air, to the discomfort of an unexpected drenching and to the danger of an unforeseen attack. All the time, whatever happens, he dare not move, lest the animal take alarm and charge, either wounding the photographer or ruining the negative.

Of all India's animals the porcupine is the most elusive, and photographing him is the trickiest of tasks. It's impossible to guess what he will do. Watch for him by the hour, and, in one way or another, he will disappoint you. He may stay in his burrow altogether, or he may pop out like a jack-in-the-box from an exit at the other end. If, by some lucky chance, he should emerge at the expected spot, he comes out like a miniature catapult, so fast that not even a two-hundredth-of-a-second exposure will catch his likeness. Under any and all circumstances, he is an exasperating creature.

Our own porcupine is quite a different creature, both in appearance and habits. His quills are from half an inch to three or four inches long, and from afar not very noticeable among his hair. He sometimes retreats into a temporary home in a low tree-hollow or in a ready-

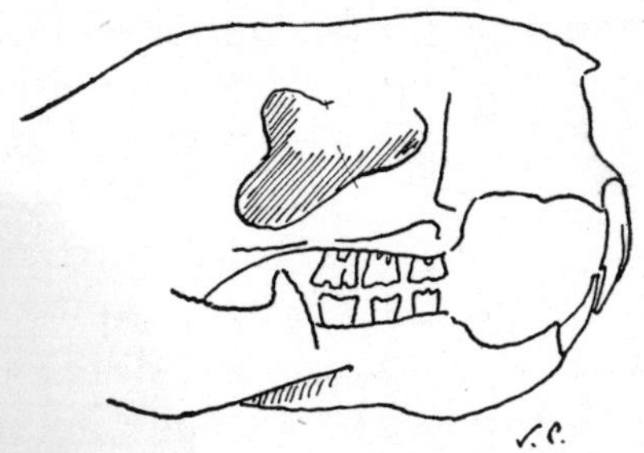

SKULL OF PORCUPINE

made cave in the rocks, or under a convenient heap of brush, but more usually he is content to be homeless, staying in one tree until he has gnawed all the bark within reach, and then simply transferring his address to a neighboring tree. Of all our forest creatures, he has the most unpleasant reputation. As compared with other rodents, he is a disgrace to his family. The silky rabbit, the diligent beaver, the provident squirrel, even the nimble mouse, all are credited with speed and intelligence, but the porcupine's virtues are nil. He is slow in motion and slow in wit. He drags over the ground; he crawls at the slowest possible pace up and down his tree. He moves only when he must. He makes no effort to build himself a house, he lays up no store of food, he endures the cold rather than take the trouble to arrange a shelter for the winter. Our campers and timbermen have neither liking nor respect for him, and consider him the laziest, stupidest and most pernicious of nuisances. Their favorite name for him is "the dunce of the woods."

In the forests along the Rockies, from Arizona to Alaska, this unpopular little beast is known as the "yellow-haired" porcupine, because of the tawny streaks that interrupt his grayish-brown hair. Elsewhere in North America, from the wilds of Canada south of the Hudson Bay region to the border of Mexico, and all through our eastern woods, he is known as the Canada, or North American, porcupine. Whether in the hills or in low country, his haunts are among the trees, but he is sometimes found in the Far North beyond the height of timber line, where he manages to keep alive on brush.

Spruce and hemlock are his favorite trees, but jackpine or aspen will do, or any other whose bark is tender. Young trees please him most, and a single greedy porcupine can destroy a hundred saplings in one winter. The lumberjacks resent his existence on this account, and some of them say he is unnecessarily destructive and takes a vicious pleasure in gnawing part way around a young tree and then leaving it for the next, just for the sake of doing as much damage as he can. But where a tree is so partially girdled, it is usually because it is too thick for the porcupine to reach all the way around from the crotch in which he is stationed. He may spend days, and even weeks, in the same tree, eating and sleeping in the same perch.

The porcupine's feet are bare-soled, and when he walks he sets them down much as we do ours, sole and heel touching the ground, contrary to the manner of most animals. The hind feet have five toes, and the forefeet only four, but there is an outer pad that takes the place of the thumb, so that he clutches the inner curve of the branch by curling his four claws around it and gripping for dear life with the pad. So anchored, he chisels away with his long, bright teeth as far around as he can.

He varies his menu of bark by an occasional attack on a young twig, drawing it close toward his mouth and snipping off the crisp shoots and leaves. In summer he may even leave his tree for a little cruise along the river-bank. He relishes the lily pads and the arrow-headed leaves that grow along the forest streams, but he rarely bothers to go into the water after them. Where they cling to the mudbank, he will stretch out a paw and grab them, but he is not inclined to exert himself an iota more than is necessary.

Any porcupine is only too ready to take to the water if it offers a chance of escape from danger, but even at such a crucial time he uses only the minimum of energy. His hollow quills help to buoy him up, and he is willing to drift along with no more than an occasional foot stroke and a little push of his powerful tail.

The one motive that makes him really ambitious and energetic is the promise of salt in the air. Every porcupine loves salt, loves it with such craving that he will risk his life over and over again to get a lick of it. He needs it, because his vegetable diet does not supply it. If he happens to come across the shed antler of some deer in the woods, he will gnaw it greedily to extract its saline content. He will literally eat mud if it has a briny taste. Wherever there is a camp or a cabin in his neighborhood, it is sure to be visited and revisited by "the dunce of the woods." He comes during the night, and in his eagerness to get at the salt, throws caution to the winds.

Porcupines can be infuriatingly noisy with their crunching and chattering and thin growls. They have an odd little squeal, high-pitched and very much like the crying of a child. The salt-seeker is anything but silent. The exasperated camper, awakened by his unceasing racket, tries shooing him away with shouts, with missiles and finally with his gun. The porcupine may be frightened off, but he invariably comes back. As long as he can crawl, he makes for the salt lure, and even with a bullet inside him, he has been known to reappear on the scene. All he is aware of is the tantalizing whiff of the pork barrel, the smell of bacon or butter or the briny wood of an empty box that once held salt rations.

No experienced man of the woods will ever leave his canoe paddle on shore, because he knows that he is likely to find it chewed to a fringe in the morning. After a few hours of paddling the sweat of his hands has soaked into the handle, and that is more than enough to bring a nearby porcupine to the premises. Experienced campers often initiate a tenderfoot, especially if they have some grudge against him, by pouring a pail of salt water over his paddle or over the walls of his canoe, knowing that a tempted porcupine will make the most of such an opportunity to indulge his passion for brine. In the morning the old-timers are full of sympathy and surprise at the shredded state of the canoe, and finally solve the riddle by remarking casually that a porcupine must have passed that way. If they think the tenderfoot has been sufficiently chastened, one of them will drop a hint that it would be just as well in the future to plant the paddle a short way out in the stream, with the handle end sunk in the mud, and to make sure that nothing of a salty flavor is accidentally left lying around in or near the canoe.

The porcupine is not given to companionship with his fellows. He travels alone, when he does travel, and within an area of twenty or thirty trees there is likely to be only a single resident. The mother finds a hollow nook in which her infants, usually from two to four in number, can be secure until they are able to waddle after her. The babies come into the world supplied with quills. They grow quickly, and when their orange teeth are of gnawing length and strength, she chaperons them for a few weeks and leaves them to shift for themselves. The youngsters may remain together for a little while, but the group soon breaks up and each one goes his own independent way.

The "dunce of the woods" has one redeeming feature, which even his severest critics admit. He is a plucky fighter. He never seeks a quarrel, but under attack he shows the utmost grit, and he can defeat creatures several times his size. A hungry fox or lynx stealing toward him sends every quill upright. When cornered, the porcupine turns tail—but not in flight. He is no coward. He turns his back on the enemy because this is his fighting position. The first step in his defensive tactics is to seek shelter for his head, which is not fully protected by quills. If there is any crevice in the rocks or among the tree-roots, he pokes his head into it. If there is no such refuge, he hides his head between his forepaws. Back arched, feet firm on the ground, quills bristling, he awaits the attack.

His enemy finds himself facing a furiously lashing tail, whose violent strokes deal terrible punishment. Every approach is cut off. The attacker, trying for a chance to get his teeth into the porcupine at some vulnerable spot, is met by the piercing quills, each one barbed and needle-sharp. A single porcupine can hold off a number of dogs by this negative style of battle. As many as six hundred quills were found implanted in the head of one of a pack after an unsuccessful engagement. Even after the battle is over, the porcupine has the best of it, for the quills remain imbedded in the body of the attacker, and his efforts to rid himself of them are in vain. Many a dog-lover has had to remove these short daggers, and even the tenderest hands cannot draw them out without causing excruciating pain.

The porcupine has a widespread fame as a sharpshooter, on the theory that he is able to aim his quills wherever he wishes and actually throw them, but he is not quite as clever as this. In the wild lashing of his tail, as he misses a blow, a few quills may fly out now and then, but he certainly cannot control their flight. He may even wound himself in this manner, but these are only accidents of war. He is not a marksman, and cannot shoot.

In some parts of America, the porcupine is confused with the hedgehog, and is so called. The hedgehog lives in Europe, Asia and Africa, but not in this country. He is not a rodent, but an insect-eater. He hibernates in winter, whereas the porcupine does not lie up for the cold weather. Both, however, have sharp spines. The hedgehog's fighting technique consists of rolling himself up into a prickly ball, so that it is impossible for any creature to get a grip on him. By way of experiment, people have tried prodding a captive porcupine to see if he would imitate the hedgehog in this respect, but he always keeps to his own manner of defense. Any creature with such a quiverful of arrows on his person needs no other. Mr. G. G. Goodwin, of the American Museum of Natural History, and Mr. Ernest Thompson Seton have taken the trouble to count the quills in a

THE PORCUPINE LIVES CHIEFLY ON TREE BARK

Canadian porcupine of average size, and report the results of their arithmetic as follows:

1,250 in the head
33,600 in the body
1,600 in the tail

Underneath this forbidding covering, the porcupine has a well-padded body, the flesh of which has furnished many a camper his supper. In most places there are rules against the wanton destruction of porcupines, except when the hunter finds himself without sufficient food.

Every race has its heroes, sung and unsung, and there is a noted little porcupine on the shores of White Fish Lake, Michigan, who has won the respect so universally denied to all his brothers. He happens to be an albino, and he is both deaf and blind. Though his light color makes him so conspicuous and the double handicap in sight and hearing exposes him to danger without the usual warnings, he has managed to escape unhurt for six successive years. His survival testifies to the porcupine's hardihood and spunk, for it is considered unlikely that any other small animal of the woods could keep alive with such odds against him. This little individual, at least, is not a dunce!

7

THE ELEPHANT

WHEN Abraham Lincoln took up his duties as President, he found among the mass of State documents awaiting his attention a formal communication from the King of Siam. This letter offered to present several elephants to the United States. His Majesty had learned from an American naval officer visiting at his court that there were no elephants in our forests. He could not understand how any nation could prosper without them. He suggested that if the beasts were turned loose in the hot luxuriant woods, where they might have plenty of vegetation and water, they would thrive, and in time their descendants could be captured and tamed and put to work.

In the East, the elephant has been the servant of man since the remotest days of history, building his roads, his temples and palaces, hauling his timber from the heart of the forest to the shore of the floating stream, transporting travelers, serving as mounts in tiger hunts, treading solemnly in royal and religious processions, performing the thousand and one tasks necessary to a population of millions, and adding to the pomp and splendor of their princes. To India, Burma, Siam and to neighboring countries of Malaysia, and on the islands of Ceylon and Borneo, the elephant has meant what the horse and the ox, in earlier days, and railroads and machinery and automobiles to-day, mean to us.

He is enormously strong. He is profoundly patient. And, most of all, he is intelligent. Six months of discipline transforms a wild elephant into a gentle and reliable slave. He learns his lessons through the cruelty of hot irons and sharp goads, but once he has been properly trained, there is no need for harsh treatment. He responds readily to a whisper, to a signal, to the merest pressure of his driver's hand. In the Orient, elephant power and elephant character are of prime importance to man.

In Africa, curiously enough, they have no practical significance at all. When the blacks of that continent first heard that elsewhere wild elephants were taken from the forest and enslaved, they laughed aloud. To them, the story was just one more "of the white man's lies." The huge creatures roamed in their dense forests, tramped across their open plains, trekked through the bamboo jungles as far up as timber line, to a height of 12,000 feet, and even in the swampy regions beyond. There is also a small variety of tusker, known as the "pygmy" elephant, in the hot, moist tangle of the Congo forest. But nowhere in all Africa had any one ever dreamed of taking the elephant captive! Dig a great pit, cover it with branches and leaves, so that he might crash into its depths—bring him down with arrow or gun—yes. Fallen, they could hack out his precious tusks, and deal with the traders eager for ivory. At the same time, they could whack the enormous carcass into thick steaks, cook the juicy trunk into a mess of savory soup, boil down the pad of the foot into lard. A single elephant makes a magnificent barbecue, enough for a hundred hunters. Singing, dancing and shouting they crowd around the vast body. Now and then one of the men steps right into the carcass to carve out some choice bit. Eat his flesh, sell his ivories—this has been the way of their fathers as far back as tradition tells. But turn the "lord of the forest" into a slave to serve the will of man—it could not be! Why, a single step of the giant foot, and

AFRICAN FOREST GIANTS

a man would be crushed to a jelly. A single sweep of the mobile trunk, and a man would be picked up like a blade of grass, sent spinning through the air, to be dashed to the ground twenty feet away. The Africans respect the elephant as not only the strongest, but the wisest of all animals. One tribe pays him the homage of worship, and when they kill one of the grand creatures, they cry out, "Pardon me!" How could puny man even imagine that he could conquer a whole herd of the towering beasts and bend their might and their wild spirit to his will?

It has been generally thought, even outside of Africa, that the African elephant could not be tamed. The reason was supposed to be that he is much fiercer in disposition than the Asiatic, and that his intelligence is not keen enough to grasp the lessons that an elephant must learn before he can be useful. But this difference in their character and intelligence has been disproved. In 1899 King Leopold of Belgium imported several domesticated Indian elephants and their keepers into Africa in order to experiment with the taming of African elephants, and the results show that the two are equally submissive and capable of training.

The truth seems to be that it is the men of the two countries that are unlike, not the elephants. Africans are simple, tribal people. A cluster of mud huts, surrounded by stakes to keep off prowling lions and leopards, makes up one of their native villages. Asia's civilization is ancient. Savage Africa's is only beginning. The blacks have no cities of their own, no buildings, no developed commerce and culture, none of the complicated activities that have occupied the Orient these thousands of years. They have never needed the elephant as a worker.

"Dead elephant" is one of the African names for ivory, and it is for the sake of his tusks that they have followed him year after year, century after century, through the maze of swamp and jungle in which he roams.

From prehistoric times to the present day, ivory has been treasured and coveted as a luxury and a tool to man. The Trojans wore buckles and pins fashioned from the elephant's tusk, and adorned their war chariots with bits of it. King Solomon sat on a throne of ivory as he delivered his wise judgments. The Greeks cut it into statues of their gods; the Romans honored illustrious men with handsomely chiseled writing tablets and scepters wrought in ivory. King Charlemagne whiled away many a long evening in his buttressed palace playing the noble game of chess with a set of carved ivory pawns, presented

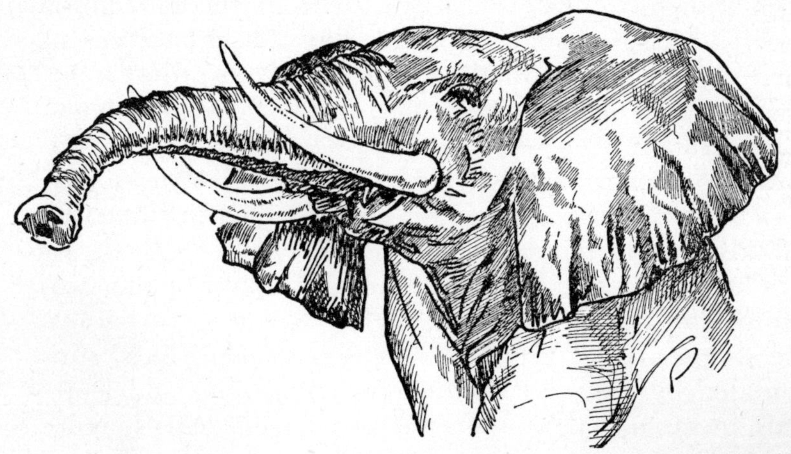

AN AFRICAN TUSKER

to him by Haroun-al-Rashid, Khalif of the Saracens. Queen Victoria rested her royal feet on a footstool made of the precious white tusk. And every one of us has known the touch of cool, gleaming ivory.

Since the beginnings of commerce, ships have headed their prows toward Africa's coasts in search of this white cargo. In times when the ivory trade is at its height, the United States alone buys a million and a half dollars' worth in a single year. England, Germany, China and even India import it in huge quantities. This vast amount of ivory has meant the death of sixty thousand elephants a year.

The captive elephants of the East supply some of this white treasure, for from time to time their tusks are cut at the tip, where they are in their hardest state. But the bulk of the world's ivory comes from Africa. Her elephants surpass all others in the length and strength and quality of their tusks. Besides, all African elephants, both bull and cow, carry well-developed tusks, whereas in India, only the bull grows them to a conspicuous degree. Even then, they are sometimes too short to be visible. Ceylon elephants generally have nothing at all to show in the way of tusks. When, once in a while, a bull is captured with a moderate pair, the people of the island regard him as something of a curiosity.

The tusks of the elephant are really his upper front teeth. Yet, spectacular and powerful as they are, they are no use whatever in biting! Anchored in deep sockets high in the skull, they project beyond his mouth on either side of his trunk, measuring as much as nine feet from lip to tip. But this is only part of their full length. A foot and a half or two feet of tusk is concealed within the head.

In Africa the natives say that every elephant is right- or left-tusked, just as we are right- or left-handed. One always works harder than the other. There is invariably a difference in their weight of from two to twenty pounds. Ivory in the raw is not the lovely, shining substance that we know in art museums and shops. Freshly taken tusks are a dirty yellow, a dingy brown, sometimes quite black with the grit and slime of a lifetime in the forest. They are coated with a rim of discolored surface to a depth of an eighth of an inch, which tells the story of their rough usage.

The Arabs call the elephant's tusk *Hadam,* which means "servant," and the name is well earned. The tusks are his excavating and scraping implements. He uses them strenuously in digging up stubborn trees, in gouging the tannin-flavored bark on which he chews with relish. Elephants have been known to toil at well-digging to a depth of several feet in dry country. The tusks make murderous fighting weapons. Terrible battles between rival bulls often take place, and the heavy, slashing tusks deal deadly wounds. Such a pair of sabers may weigh around two hundred pounds apiece!

A handsome pair of ivories is a doubtful blessing to an elephant, for hunters are constantly on the watch for tusks of phenomenal size. They vary greatly, even among the largest of elephants. They may weigh as little as thirty pounds apiece, or as much as eight times that. The tusks of the African cow may be as long as those of the bull, but they are lighter in weight and softer in texture, and consequently are of less value as ivory.

Of all the African tribes that follow the trail of the tusker, none is more expert than the Wandorobo. Familiar as they are with his haunts and his habits, they have no name for him other than *ol janitu sabug*—"the big thing." Suitable enough for an animal that stands eleven feet high at the shoulder and weighs between five and six tons! An African bull elephant would outbalance three heavy automobiles. His skin alone tips a tannery scales to a thousand pounds.

If one of the Wandorobo, or any other African, ever chanced to see an Asiatic elephant, he would be vastly surprised.

"The big thing" of the East would appear to him undersized, badly formed, and quite wrong in a number of ways. The Asiatic elephant weighs a ton or two less than the African. He is a foot or two shorter at the shoulder. His head clearly shows the outline of the skull with its bumps and hollows. His back is decidedly arched, whereas the African elephant's back dips in a pronounced concave curve. There are other minor differences, such as the size of the eye, and details in their trunks, teeth and toenails.

IVORY

Size, head, back, eye, trunk, tusks and teeth all distinguish the two, but the one infallible and unmistakable clue to an elephant's ancestry is his ears. Those of the Asiatic elephant are pointed at the lower tips. They look oddly like the map of India in outline. Though they may be twenty inches long, they are tiny compared with those of the African. His are huge flaps. They average four feet across, and on a particularly huge tusker they may even measure six feet! An explorer of early days tells how comfortably he slept once, when on a hunting trip, on a mattress improvised from a pair of elephant ears.

Every one who has struggled to master the accomplishment of wriggling his ears has reason to envy the elephant. He can move his enormous flaps backward and forward with the utmost ease. This is more than an accomplishment in the case of an elephant. In the intense heat of the tropics they are a great convenience, for he uses them as fans! They also perform the service of driving off the swarms of insects that disturb his peace, pricking and prying for a chance to feed on his head and body.

When his suspicion is aroused, an elephant does not "prick up" his ears. He unfurls them, spreading them sidewise, and they stand out like a pair of thick, rough-edged, flat umbrellas, giving the African elephant's head-on view a width of ten feet from ear-tip to ear-tip. In the years of friction against the rasping trees and all the rough treatment they get in his forest expeditions, they become tattered and frayed, ragged and jagged, scarred and torn. They are sometimes punctured with gaping holes. As the elephant grows out of the calf stage, his ears turn over at the top rim in a backward roll. The inner ear is very sensitive, and the elephant has an extremely acute hearing.

But it is on his sense of smell that he relies chiefly to bring him news. The elephant's nose is developed to a length of eight feet, in the form of his strong and supple trunk. It has only to brush lightly over the ground to catch the scent of friend or foe; it has only to swing through the air to catch the messages of the wind.

This proboscis, however, is much more than a nose.

Surrounding the two valves that run through it, is a maze of interlaced muscles whose interplay provides both strength and dexterity. The number of these muscles has been estimated at more than 40,000. With his

AS THE ELEPHANT GROWS OLDER HIS EARS TURN OVER AT THE TOP

trunk, the elephant breathes; he feeds; he drinks; he fetches and carries; and even bathes himself.

His short, stocky neck furnishes a firm support for the bulky weight of his head and the burden of his tusks, but it has no stretching power, and gives him no help in reaching his food or drink. Without his trunk "the big thing" would die of starvation and thirst. It mows the grass for him; it strips the leaves from the branches; it conquers resisting trees. At its tip are two little finger-like tabs, which he can pinch together like a pair of tweezers. He can probe into the heart of a young plant and scoop out its tenderest kernel. He can pick up tiny bits of grass and seeds from the ground as deftly and firmly as we grasp small objects between our finger and thumb. In the African elephant, these two extensions are about equal in length; in the Asiatic, the one on the upper margin is longer than the other. They are acutely sensitive, and can locate fragments as fine as a sliver on the ground.

When the elephant drinks, he sucks up a stream of water into his trunk and holds it stored in the valves, curling the trunk up to pour it into his mouth. When he bathes, he uses the shower bath method, squirting a spray of water through his trunk over his back and shoulders. He can even wash behind his ears in this way! On a hot journey, far from water, he often refreshes himself with a dust bath, which he likewise manages through the pumping and blowing powers of his trunk. He has even discovered that he can break off a leafy branch with it and wave it back and forth effectively as a fan! In Sanskrit, the elephant's name is *hantsin*—"the animal having a hand."

As though he realized that his trunk means life and death to him, he protects it carefully. He seems to know just how far to go in using it. By curling it up and releasing it suddenly, he can deal a heavy blow with it, and he does do some of his milder chastising "by hand." But in hard fighting, he relies on the stabbing force of his tusks and the crushing power of his feet, curling his trunk up to keep it out of reach.

Elephants live from eighty to a hundred and fifty years, and every day of their long existence has its own share of change and adventure. They are gypsies, forever on the march. Like gypsies, they have their routes and times of travel. From year to year, from season to season, from day to day they scuffle through the world of trees. Their wanderings are not aimless.

During the season of the heavy rains, when a chilly cascade drips from the trees, they find the forest uncomfortable and take the trail to more open country. There they cruise about, finding shelter from the lashing rains under a convenient clump of flat-topped acacias and enjoying the warmth of an occasional hour of sun. When the rains are over, and the heat on the plains grows too intense, they trek back to the cool shade of the forest again.

As surely as country boys make for the hills and woods in berrying and nutting season, elephants head for the wild-fruit groves. They seem to sense the picking season from the fragrance of the air. Theirs is a sweet tooth. They love the pungent, juicy fruits of the tropics. Their pilgrimages from one wild orchard to another take them hundreds of miles, but the reward of such dainties as plums, breadfruit, wood-apple and the luscious pineapple makes the journey worth while. As long as there are fruits for the plucking, the huge beasts linger, and when the last branch is stripped bare, they return to their ordinary fare.

They are heavy feeders. Their giant jaws are never still. The Asiatic elephant lives largely on grass. The African shows a preference for the leaves and shoots of the trees, particularly the mimosas, but both relish wild cabbages, plantain, young palm and fine bamboo. Elephants love variety in their food, but their principal need is quantity. It takes half a ton of vegetation a day to still the appetite of one of them.

On their feeding forays, the big herds break up into small bands. They seem to know that where a few can

INDIAN ELEPHANT

make a rich harvest, the full herd, two or three hundred strong, would starve. All day long, and most of the night, their long trunks reach out for provender and cram great masses of green into their pointed mouths. Their huge cheek teeth, each one fully seven inches wide, grind and mash the crisp herbage violently. From the time the elephant is born until he reaches the age of sixty, he is constantly growing new teeth. It is not his tusks that keep him in the "teething stage" for over half a century. As a baby, he sheds his milk tusks, and the pair that replaces them is permanent. But his grinders have a more complicated history.

The business of crushing and chewing that goes on steadily day after day wears their surfaces into irregular grooves, and under this strain they become deeply channeled. At the ages of two, six, nine, twenty-five and finally at sixty, when the elephant's life is half over, new molars push their way along the curving arc of his jaws, and crowd out the old, worn teeth.

Wherever an elephant herd has passed, it leaves a scene of wreckage. The ground is trampled flat; the grass is crushed; uprooted trees are scattered along the route. Unable to reach the spreading branches, they attack the tree at the base. Trunk meets trunk, and in the tug of war the tree is jerked up by the roots. Even trees of thirty inch diameter yield to the pull of the elephant's "hand."

As they travel through the dense forest, they usually march in single file. It is a slow procession, with stops for feeding, and detours toward a salt lick or toward water.

Here and there a tree is left standing. These solitary trees have their bark rubbed smooth by the great beasts as they scratch their itching sides against them. Some of them are worn high above the shoulder line, and for a long time this was a cause of wonder. It is now known to be due to a particular cause.

In the elephant's head, between the eye and the ear, there is a tiny deep pore, which at times becomes clogged up with its own moist secretion. It irritates the elephant, and he stops and scrapes it off by rubbing his head against the tree. The story goes, among some of the natives of elephant countries, that the big beasts have sense and skill enough, when they feel this irritation, to pick up a tiny twig between the finger ends of their trunks and insert it into the clogged pore, to clean it. Often, when an elephant is brought down, the men will crowd around his head fighting and squabbling for first chance at finding the "little stick," which they cherish as a good-luck piece. Such bits of twig are sometimes found, but they are only fragments of bark or twig that have gotten lodged inside during the rubbing. No one has ever seen an elephant performing this intricate operation. The captive beasts have this little detail taken care of for them by their keepers.

Elephant trackers know from the condition of the vegetation how recently a herd has passed and which way it has gone. As the animals are protected in practically all districts, each hunter works on a limited permit, and wants to be sure that the herd contains one big enough to satisfy his ambition. He can tell almost to an inch how large a tusker he is following, simply by measuring the footprints. By an odd coincidence, it happens that the circumference of an elephant's forefoot is just about one half his height. If the tape measure runs sixty-six inches around the imprint of the forefoot, for instance, the hunter knows that his elephant will stand eleven feet high at the shoulder.

From the external appearance of the foot one might think the elephant flat-footed. It is hard to realize that he actually goes through life on tiptoe! The five toes of his foot form an arch, which is filled in and supported by a tremendous pad of muscle and fat. These arches must support a weight of five tons. With every step, this great weight makes the elastic pad spread, and as the foot is raised, the pad contracts. The huge beast literally walks on cushioned feet.

Although each foot is five-toed, only the front feet

have five toenails. The African elephant's hind feet show only three toenails, and the Asiatic, four—that is, as a rule. There are occasional exceptions. In India, when a man is buying an elephant, one of the main points he considers before making his purchase, is how the toenails add up. If the sum comes to an uneven number, he hesitates, as there is an old superstition that

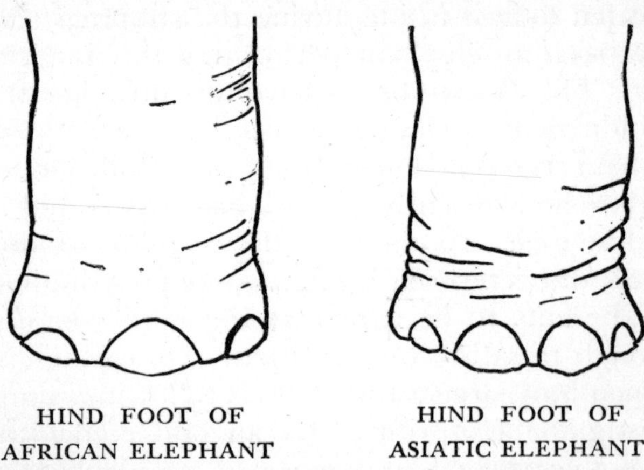

HIND FOOT OF AFRICAN ELEPHANT

HIND FOOT OF ASIATIC ELEPHANT

odd numbers in an elephant's toenails bring bad luck.

Huge-footed and huge-bodied as they are, the great pachyderms are not clumsy. They climb sure-footedly, and in making a steep descent, which is even harder for such heavyweights, they feel their way skillfully. They will test the firmness of a sandy hill by pressing their trunks against the soil, and where the footing is uncertain, they manage by sliding until they reach ground that will support them.

Even in a sudden stampede, they avoid collision. When a calf is frightened, he takes refuge under the lee of his mother's body, standing far up forward near her forelegs. It is an old habit of his babyhood, for there he got his first nourishment, suckling at her breasts, with his trunk curled up and out of the way. The mother is very careful, even in the haste and terror of flight, to avoid trampling and crushing him.

Baby elephants need a touch of discipline now and then and they get it from their mothers. An unruly calf that tries to run away, or becomes bothersome, is apt to feel the sting of his mother's trunk. She can curl it up and let it fly out like a released spring, and such a blow carries no small amount of force. An occasional spanking brings the calf to his senses, and after one of these hidings, he sticks more closely to the rules of elephant behavior.

The trunk that chastises also caresses. The mother strokes the little body with it, draws it close to her in an embrace, and where the going is rough on the road, she boosts the helpless little calf along with a shove of her trunk.

It is not only the babies of the herd that look to the cows for guidance. The oldest and mightiest bulls look to her for leadership, as well. It is she who heads the line of march, who keeps a lookout for danger, who summons the herd for massed action. It is she who gives the shrill scream that warns them. Elephants trust to the wisdom of the mothers in the herd, and they follow the orders of their matriarchs. In the charge the cows are in the front ranks; in the actual fighting the big bulls crowd forward and assume the brunt of battle.

Elephants are a noisy crew. They gurgle while they eat, they snort, they whoosh and whine with pleasure over the succulent green in their mouths. They quarrel over a favorite feeding spot, grumbling in protest as one shoves the other aside. The rhythmic flapping of their ears adds to the medley of noises. The babies squeal, the grown-ups grunt in their throats or rap their trunks with a nervous tom-tom against the ground or the trees. Not the least of their noises is the constant rumbling that goes on in their insides. It can be heard a hundred yards away, and sometimes betrays their pres-

A TOUCH OF DISCIPLINE

ence to the hunter when they are unseen among the trees. The scattered groups usually contain a number of cows, a few calves, and the younger bulls. The older tuskers often withdraw for a touch of solitude. The Africans say that the rumpus raised by the family is too much for the nerves of the father, and that he wanders off from time to time for the sake of peace.

Elephants have one quality that sets them apart from

all other herd animals. They are capable of loyalty of the highest degree. When one of their number is wounded, it is not uncommon for several of his comrades in the herd to cluster around the fallen giant and combine their strength to help him to his feet and lead him away to safety. Bracing his huge, helpless body with their trunks, lifting him with their tusks, they lend every ounce of their might to his rescue. Usually such devoted help comes from the cows of the herd. The most hardened hunters confess that they are moved to admiration by such heroism and unselfishness.

The law of the herd can be severe, as well as kind, among elephants. It frequently happens that a bull spends most of his life alone. No one knows why, nor how it comes about, but from time to time some powerful tusker withdraws from the company of his fellows. Whether he is banished, or whether he voluntarily chooses a solitary existence, is uncertain. But for a century or more, such a lone bull makes his way unfriended, unmated, an exile from his kind. Either the herd makes him feel, through some silent verdict, that he is unwelcome, or else his own morose nature unfits him for their companionship.

These bachelor elephants sometimes turn ugly and go wild in destructive fits of violence. They are known as "rogues." When such an animal ranges near the habitations of man, he is a fearful menace. Plantations are wrecked, houses torn down. A chaos of terror clutches the countryside. There is not a moment's security, until the beast is destroyed. The villagers unite in bands to hunt him down, and hunting a "rogue" is a wild and dangerous undertaking.

Tame bulls are likewise subject to sudden fits of frenzy, and their keepers are constantly on the watch for symptoms of restlessness in their animals. They can generally foresee the coming of such an attack of temporary madness, though the outbreak may occur without warning. The gentlest elephant in the world is likely to "go musth," as they say in India. He may recover from the spasm, but he is highly dangerous and must be heavily shackled and kept apart from the others in the corral. Sometimes the frenzy is permanent, and then he must, of course, be killed.

It was the threat of such a collapse that brought the famous elephant Jumbo to America. He had for years been the center of attraction at the London Zoo. Jumbo was an African elephant, gigantic in body and gentle in behavior. He was one of the most renowned and beloved personalities of his day! People flocked in thousands for a view of him. Every child in London longed for the thrill of a ride on Jumbo's back. Suddenly the news broke that Jumbo had been sold. He had been showing signs of nervous irritability, and the authorities at the Zoo felt that he might one day go wild. When P. T. Barnum, the American showman, put in a bid of ten thousand dollars for Jumbo, the magnificent, his offer was accepted and London's dearest pet was shipped across the Atlantic. He recovered from his fidgets, and had a sensational career in this country. The English papers were filled with letters, many of them from heartbroken children, clamoring for his return, but Barnum resisted all entreaties. Jumbo traveled in the Barnum circus, and was killed by a freight train as he was crossing the tracks of a railroad, on his way to his box car.

Elephants in zoos and circuses need expert care. Sometimes, men are imported from India to look after them. The Indian *mahouts,* or elephant drivers and trainers, are natives who have inherited their calling through generations. They are expert not only in the care and control of the beasts, but adept at the highly dangerous *Keddah*—the art of tracking and trapping the wild herds and driving them into captivity.

Although the practice of using elephants for every conceivable kind of heavy work dates back to India's earliest history, they have never been bred in captivity, like the beasts of burden of other countries. The calves are not good for service until they are ten years old, and the expense of feeding their enormous appetites, as well as taking care of them for this long period of uselessness, makes it unwise to breed them. Besides, for some strange reason, an elephant that has been taken wild and tamed is more obedient than one born and reared in captivity for working purposes.

The ancient *Keddah* system is still followed in the East, although less extensively than formerly. Whenever fresh recruits are needed for the working herds, the leader of the hunt summons several hundred skilled *mahouts* and beaters, and they ride out to the hills, mounted on their tamed elephants. All of them, men and beasts, have had special training in the *Keddah*. When the advance scouts report the presence of a wild herd in the neighborhood, the whole outfit makes for that region and surrounds the district where they roam. Gradually, the beaters narrow their circle, hemming in the wild herd closer and closer. Meanwhile, a strong stockade has been built of stout posts interwoven with tough bamboo. The forest echoes with the screaming and trumpeting of the elephants as they crash through the trees in panic. The men flash lighted torches, shoot off guns and rockets, beat their drums and shout and yell, to confuse and terrorize the beasts, driving them steadily toward the stockade. This is the *Keddah* proper (the word itself means "fence"). With all escape shut off the desperate animals plunge through the gaps of

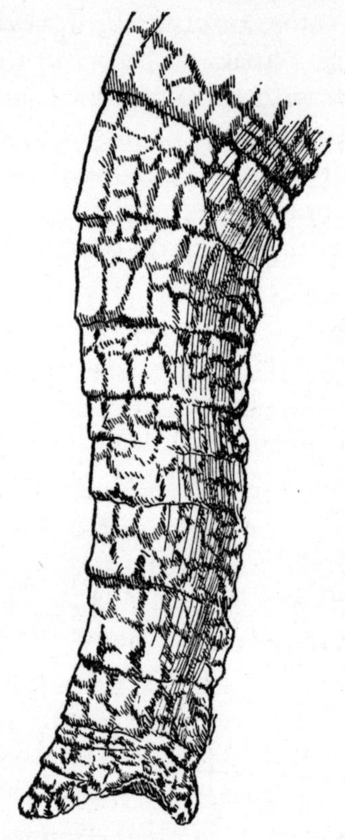

DETAILED STUDY
OF TRUNK

the stockade. The heavy gates bang shut behind them—and the herd is trapped.

Perched high on the backs of their tame mounts, the *mahouts* and noosers ride amongst the captives. They are armed with ropes and chains. One by one, the noosers slip to the ground, watching for a chance to steal up to the wild beasts and shackle them. It is terribly dangerous work, and without the astute help of the tame elephants, it would be impossible. They do their duty as disciplinarians, forcing the rebellious captives against the wall of the stockade, beating them into submission, striking them with their trunks, stabbing them with their tusks, crowding and pushing and wounding them without mercy, until the chains hold them fast.

On the march to the training camp, each new captive tramps the road under the escort of two tame elephants. He is chained between them, and their combined strength exhausts him as he struggles to break away. The captive is led to the training camp and subjected to a six months' course of school. He learns to kneel, to move cautiously, to submit to orders and to work. Fear makes him gentle, and his remarkable intelligence does the rest.

Indians recognize three "castes" of elephants, and at the fairs held for the buying and selling of tamed elephants it is the custom for the purchaser to state in advance which of them he wants. Nowadays these fairs are less frequent than in the days when all the work of the country was done by elephant power, but they are still conducted on the old methods. It takes days to close a sale. The elephants are set up in stalls, paraded back and forth, put through their accomplishments and examined from trunk to tail. The *Koomeriah* brings the highest price. He represents all that is perfect in an elephant from the Indian point of view in the way of size, strength, obedience and good looks. Even his tail must be beyond criticism, well tufted at the tip with crisp, black hairs. The *Dwasala* comes second in class, and includes animals of excellent build and power, but not up to the very highest standards in one way or another. The lowest caste is the *Meerga,* rather thin-legged, thin-trunked, and neither handsome nor particularly good-tempered. Their special virtue is their speed, and they are often put to use for travel and as message carriers in preference to the more valuable elephants.

Once in a great while an elephant is found with a pale skin mottled with a few pinkish blotches on his sides and a rosy patch or two on his ears or trunk. Then the excitement runs to a high pitch wherever the news of his discovery spreads. Usually such an elephant has eyes of a faintly yellowish color, and his toenails are milky gray. These few delicate tints raise the elephant to glory, and bring his lucky owner to fortune. The word goes out that a "white elephant" has appeared, and the whole country rejoices. In Siam the finding of a white elephant is immediately reported to the King. Since ancient days the Siamese have regarded the white elephant with reverence, and some of the people of the interior even now believe him a divine creature. When such an animal was brought to the royal court, rich presents were showered on both him and his captors. The elephant received lofty titles of rank, bestowed in public ceremonies by the King himself. He was addressed as Gem of the Sky, Glory of the Land, Radiance of the World, and his coming to court was believed a sign of favor from Heaven. He was gorgeously ornamented with trappings of silk and embroideries set with jewels, and canopies of brilliantly hued tapestry were set in frames and perched high over the august person of the adored creature to protect him from the heat of the sun. He was housed in a magnificent stable, fed and watered from silver buckets, and a cabinet of ministers of noble status was appointed to attend him and look to his wants. The privilege of fanning him was a coveted honor accorded to a man of princely descent.

White elephants appear very rarely, perhaps on the average of once in a hundred years, and several of them have featured in the making of history. In the sixteenth century a fierce war was waged between the nations of Pegu and Siam over a bitter dispute as to the ownership of one of the treasured animals. Five kings and many thousands of men perished in battle because of him.

During the reign of Queen Victoria there was an exchange of courtesies between England and Siam, and when the Queen's envoy returned, he brought Her Majesty a small gold casket which the Siamese king had put into his hands, charging him to carry it to her in person. The little box was unlocked with a golden key and, to the surprise of the Queen and her Court, disclosed the precious gift which Siam had sent to England—several hairs taken from the hide of one of the royal white elephants.

The Queen herself received a high compliment from one of the Siamese ambassadors, who described her in a letter most flatteringly. "One cannot but be struck with the august Queen of England," he wrote. "Her eyes, complexion and, above all, her bearing are those of a beautiful and majestic white elephant."

The arms and flag of Siam show a spotless, snowy elephant as the country's symbol, and this portrait has led to the popular belief that the famous creatures are pure white. In 1633 one was somehow obtained by Holland and displayed there, but none was ever seen this side of the Atlantic until Barnum imported "Toung Taloung." It cost Barnum over a quarter of a million dollars and untold effort to get possession of him. His representatives traveled across the Pacific and down the coast of China, ending their slow journey at Siam. They tried with offers of a huge sum of money and with every form of coaxing and argument to bargain for a white elephant, but the more they urged, the less they accomplished. The word ran around that these men had come to deprive the kingdom of one of its idolized animals, and the indignant populace threatened to attack them. Barnum's agents left the city and traveled up into Burma, where they finally succeeded in getting a white elephant. They bought the animal from the widow of a nobleman, who was unable, after her husband's death, to maintain the creature in suitable splendor. They carried on their transaction in secret, as far as they could, and managed to float their huge purchase on a boat down the Irrawaddy River as far as Rangoon, where they transshipped him and set sail toward Singapore. They congratulated themselves on the success of their venture, not only in having made the purchase, but in having escaped harm. All along the river trip, there had been danger of riots, and they were glad to be out of the country. However, they gloated too soon, for the elephant died on the voyage to Singapore. It was thought that one of the crew, or some person who had come aboard purposely, had poisoned the animal rather than permit its removal.

Barnum's men returned, crestfallen, to America, with only the story of their misfortunes to show for all the money and effort they had spent. A short time later, Barnum tried again, and with the help of some influential English residents, his men persuaded the King himself to consent to the sale. They had to give their promise, in writing and under oath, that the elephant would be as carefully tended and as respectfully treated as if he were at home. This time they traveled overland, from Mandalay to Rangoon, and the animal was most scrupulously watched throughout the journey. Angry crowds surrounded the party wherever they stopped, but the sanction of the King protected them and they came through safely.

Toung Taloung stood eight feet high, and had handsome tusks a yard in length, but he was a tremendous disappointment to the American public. People came to Barnum's circus, expecting to see a chalk-white elephant, and he was merely a pale gray, with a few pink patches on his head, trunk and ears and ivory-tinted toenails. However, Barnum made a great deal of money by exhibiting him, though one of his friends suggested that he could make more by giving him a coat of whitewash.

Wherever there is talk of elephants, the world over, sooner or later the subject of "elephant cemeteries" comes up. It is quite widely believed that far in the hidden maze of the forest, remote and secret beyond discovery, there are silent groves where the gray giants have their final home. The story goes that when an elephant feels his strength ebbing, he takes to an unseen trail and plods grimly on until he reaches the spot where his kind go to die. In some places it is even said that elephants bury their dead. These "cemeteries" are pictured as piled mountain-high with gleaming tusks—a treasure hoard of ivory waiting for the lucky hunter who shall one day stumble upon the invisible trail that only dying elephants know.

No man has ever found such a spot, but the belief in it persists. The basis of it is the apparent absence of elephant skeletons and tusks in the forests through which the animals roam. There have been men who spent most of their lives tracking elephants without ever once seeing any trace of their remains, and as they never could account for it in any other way, they were inclined to support the ancient story of the hidden cemeteries.

It is true that tusks and elephant bones are very rarely found, but to the eyes of the trained searcher the forest reveals its mysteries. When an elephant grows weak with age or illness, he is unable to keep up with his herd, and wanders about by himself. He remains near water, and sometimes rests his aching body in a shallow stream or in the cool mud along its bank. Lacking the strength to rise, he may sink to the bottom as he dies, and then, of course, his body is never seen. Or, he goes deeper and deeper into the forest, finally lying down among the thick vegetation when he can travel no more. There he drops into his last sleep, and the hungry life of the forest, the scavenger jackals and hyenas, the birds and field mice and insects dispose of the carcass rapidly. The tropic heat brings on quick decomposition, and the lush, leafy growths thicken luxuriantly over the fragments that are left. The damp soil absorbs the bones and tusks. Under the forest camouflage, the gray giant disappears.

8

THE LEMMING

Many, many years ago, there sprang up an old story about a town in northern Germany that was suddenly invaded by an army of rats. Rats scuttled through the streets and into the courtyards, and finally, becoming bolder and bolder, entered the houses and robbed the storerooms and pantries.

The villagers did everything they could think of to get rid of this pest that had descended upon them from nowhere. But all their attempts were futile, and the rat army grew larger and larger.

Finally, the mayor called a meeting of the town council to devise some way of ridding the place of the enemy. But they could think of nothing. Then the mayor offered a large reward to any one who could succeed in banishing the rats. To his amazement a stranger appeared in the town and offered to earn the reward—a ragged, rather crazy-looking man who, the mayor felt very sure, could do nothing of the sort. However, he was in despair, and he told the fantastic person to go ahead.

What was his astonishment to see the man proceed to the town gates; take from his clothing a flute and begin to blow a curious little tune upon it! And suddenly from the cellars and bins, from the houses, courtyards and lanes, came hundreds of rats, scurrying down the road leading out of the town! And out of the town, down the road, skipped the Piper, followed by the procession of rats; skipped along until they reached the banks of the Weser River. Then the Piper stood still and played his tune, while all the rats, under the spell of the strange music, ran on into the river and were drowned!

And so Hameln got rid of its rats, and all the generations of children since that time have read the story of the Pied Piper—a story which may or may not be strictly true! But to-day there is a country north of Germany from whose borders no piper has piped her plague of rats, although many of them meet a fate similar to that of the rats of Hameln.

This country is Norway, and her rats are the lemmings, her only native animal. They are little yellowish-brown, furry and bad-tempered animals who appear in disorderly armies, eating as they go.

They eat anything they can find. When a hundred or two lemmings get into a vegetable garden, they can do a lot of work there picking the farmer's vegetables and consuming them, for they do not eat meat and they are always glad to find nice tender vegetables or young grain shoots. When they cannot find these, they will eat grass, or even the bark of trees. But when they can, they invade the gardens and grain fields in hordes, and the farmers and dogs cannot get rid of them. They are particularly fond of oats, but they do great harm to any grain crop, standing or stored. If they pass through a village in winter, no stored grain is safe from their greedy persons. They appear without warning; perhaps the first sight of them will be a small band crossing the road in some small village.

If farmers were able to set many men or dogs upon them, their destruction might be more complete, although more would be sure to appear within ten years or less. However, Norway is rather unlike most other countries in the way she has been settled. The land is very mountainous and the early inhabitants soon found that a living from the sea was far more certain than one

THE LEMMING

from the small amount of land possible to cultivate. So instead of building large castles and feudal estates, with acres and acres of land and many serfs to care for it they expended their energies upon the sea.

Back in the time of the Vikings, the Norwegian overlords were bold sea pirates who spent their wealth not on building fortress castles, but in equipping the strange, huge old Viking ships and manning them to sail the ocean and capture what they could. Besides piracy and fighting, there was the natural harvest of the seas—whaling and fishing. And down through the ages, the Norwegians have been sailors and fishermen, living near their coast, and on their waters.

So the few homes that sprang up inland, with the exception of two or three cities, were very small simple farms and have stayed so—wooden cottages of a few rooms, and a few acres of land around them, usually planted in oats, barley or potatoes. There is no very great wealth in Norway even to-day. The majority of people live simply, and because they live simply and their country is one of the most sparsely settled countries in Europe, the destruction waged upon their crops by any pest is just that much greater, for their small crops are their means of living.

The lemming plague from ancient days has been one of the country's greatest menaces. Back in the sixteenth century, when the cathedral town of Trondhjem was the chief city of Norway, two archbishops of Trondhjem once made the long journey from Norway to Rome, as was the custom of church dignitaries. There, one day, they fell into conversation with a scholar from Germany. In exchanging tales of their countries, the Trondhjem bishops told the German about their terrible problem—the lemming—who appeared in such masses and without any warning. The German asked where these animals came from, and, in all seriousness, the two churchmen told him that they fell from the clouds! Nor did the German find anything out of the way in this explanation, for in those days it was quite an ordinary explanation of the otherwise inexplicable appearance of any creature. In that same way scholars had explained the appearance of small fishes on dry land after a storm. A great naturalist of the time even included in his book a picture of lemmings falling from the clouds, while below them stood ermines, waiting to seize the victims from the skies!

It was not until many years later that the belief of the Trondhjem bishops and of many other learned men, as well as of practically all the country people, was contradicted.

This is not strange, for so hidden are the ways of this small animal that you may wander for miles over places thickly populated with them, and never see one! So years went by before the mountain homes, from which they descended in hordes to invade the grain fields, were discovered.

Lemmings do not see any more of each other or any one else than they can help. They have very unpleasant dispositions. In the birch and willow zones, and sometimes as high as the snow line of the mountains and high places of Lapland, Greenland, Siberia and North America, they make their home, but it is in Norway that they are most plentiful and most unwelcome.

Unsociable and inconspicuous, darting out from under stones, or behind tussocks of grass, and generally doing most of their business at night, they live hidden lives until a migration year comes and they suddenly emerge from their hiding places.

If a lemming does encounter another lemming, or any other animal, he is immediately ready for a fight. He backs up against the nearest stone or tree trunk, sits up on his haunches, and begins proceedings by hissing or uttering a high, short bark at the enemy.

In summer the lemmings make themselves nests under rocks and bushes where they hide away during the daytime. In winter they have to tunnel under the snow to get food and protection, and here they make themselves temporary homes, carrying into the snow tunnel wisps of straw for their nests. They do not hibernate, however, and do not store food. As a consequence of shortage of food, a good many of them perish every winter.

When warmer days come, a lemming is apt to wake up suddenly to find that the walls or roof of his snow house have melted and that he himself is sitting very publicly in a puddle of melted snow, while near him several of his relatives also sit in puddles, agitated to find themselves so suddenly revealed to the world.

They seem to hate publicity, for when the snow goes, they immediately take to the bushes and rocks again. In such protected spots, the lemming female bears her numerous young in summer. In ordinary years she has about two litters, of five or six each, during one summer; in extraordinary years, she has so many children during the course of one summer that one lot are not out of the nest before another lot arrives there.

When lemming communities get too big for the food supply of the places they live in, they begin moving to other regions.

One day there will be a sound of agitated rustling in the grass; of unusual activity for daytime. Scurrying small feet dash down the mountain side. Then more and more, all through the day and long into the dark night—away down the steep mountain slopes. A small fat one will fall flat on his round stomach over a stone, roll over, pick himself up and run on, not even stopping for a fight with the two or three others he has hit while rolling. The whole mountain side seems suddenly crowded with lemmings coming from every direction, restlessly moving about. It seems as if the place had suddenly grown too small for them and some instinct had warned them to move on at once to new feeding grounds and less crowded regions. The departing animals are chiefly younger ones; when they have all left, there is space and food enough for the old ones to live upon.

There is a current impression that lemmings march in orderly ranks, like an advancing army, about three feet apart, and in regular formation. This is not true. One reason it is so hard to say anything absolutely definite about a lemming migration is that these migration marches are so utterly disorderly. The animals go any old way to get there—wherever they are going!

Unsociable and irritable as they are, they obviously prefer to march all alone, as if each one were going on a perfectly independent journey, ignoring the scores of relatives taking the very same trip! When they do meet, they bark unpleasantly and pass on, or they stage a fight. As a lemming dies at the slightest injury, and as such meetings are really unavoidable, these fights serve to decrease this marching horde considerably—a decrease which, however, is in no danger of wiping out the plentiful tribe.

From time to time down the mountain sides, hundreds of other lemmings pour out from under rocks and bushes to join the rush, until it seems as if the mysterious destination for which they are all aiming would be even more crowded than the places they are leaving.

In the wilder regions they are followed by wild enemies—wolves, foxes, Arctic owls. On the lower slopes, as they strike houses, cats and dogs run after them down the streets and people come out to watch the swarms of animals, or to chase the bolder of them from their barns and yards, saying to each other, "Another lemming year has come."

Many peasants, who have heard the tale from their

parents and grandparents, still hold fast to the belief of the old archbishops of Trondhjem, that the animals fall from the clouds.

Then, as in any animal migration, there comes a sudden decrease. Some of the country people believe that the animals left behind in the migration eat themselves to death on the plentiful green young spring shoots, but the real reason is, probably, that they cannot stand the change to a lower level. Thousands of them perish on the march; many because of the difference in climate, many due to accidents. Many are eaten by other animals or killed by irate farmers whose crops they have consumed. Thousands perish from the infectious diseases which nearly always result from too great crowding among animals.

None of those who reach their final destination return. The march, which begins with such a gathering of the clans and such a rush down the mountain slopes, slows up as the animals near the valleys. Days go by and they move on and on, but more slowly and with halts. Spring comes, and then it is easier for them to find food. Due to their independent style of marching several of them often get separated from the others, sometimes rejoining the main swarm in several hours, and sometimes not for several days or not at all. But they keep moving, and always in the same general direction—toward sea level.

In this way a small group of lemmings once attempted to put themselves on exhibit in the Museum at Oslo. Left behind by their contingent, they strayed around the city streets until they came to the high flight of granite steps leading to the main entrance of the University buildings, where the Museum is. Lemmings seem to have little natural fear of human beings. In fact, when they encounter human beings, or rather, the lower parts of human beings (which must be about all they are aware of), they show a decided inclination to attack, and give away their whereabouts by their barking cry of rage. Into a shoe, which must seem to them like an advancing continent, they sink their sharp teeth with great vigor and tenacity. However, if the owner of the shoe makes a movement of counter-attack, the valiant lemming loses his silly head completely and takes to a crazily confused flight.

Fear and excitement are said to occasionally cause the death of one of these temperamental creatures. Perhaps it is because of this hysterical disposition that they are so difficult to keep in captivity.

These visitors to Oslo, however, encountered no obstacle, and ascended the steps of the University to the very top. There they headed in the direction of the Natural History Museum. But they never got inside, because a quite unwitting reception committee met them at the doorway—one surprised scientist, quite unaccustomed to having his exhibits walk in of their own accord, and at whose surprised start they turned and scuttled down the steps again.

Whether lemmings are particularly stupid, or whether wandering confuses what sense they may have, it remains a fact that their habits on the march, except for securing food, do not seem very bright!

If they encounter an obstacle like a hay stack, instead of going around it, they make every effort to go through it! If they come to a cleft in a rock, or a "jumping-off place," they go straight ahead and jump off—into space.

In their every-day life they show no particular desire to go into water, and heavy rain storms always leave many dead lemmings in their wake, but on the march, these animals continue their policy of marching straight onward, and dive right into the icy, foaming mountain streams or deep wide lakes. One river, the Leirungen, is glacial in origin and in temperature, and very rapid. Dogs do not at all like to face it. But the lemmings cross it in thousands. This means the end of some of them, but the survivors scramble out again on the other shore, sleek and dripping.

These animals may march for a year or two before they stop—if the tragic end of this blind migration may be called a stop.

This end is a body of water called the Atlantic Ocean. Over the rocks and shores the swarms of little animals advance, drawn into the water by some instinct apparently as fatal as the music of the Pied Piper.

Into the water they go, and begin to swim out! Waves hit them; whaling boats and fishing smacks come toward them pushing in among the masses of bobbing little wet heads. Lemmings swim high, half the body and the head above water, and they do not seem able to dive down at approaching danger. When disturbed by a boat, they become confused and swim around and around in circles.

From this there is no return migration. They have come to the end. The Atlantic Ocean is very large and they are very small. They never turn back. The same instinct that first started them down the mountain side seems to urge them ahead.

This fatal end has been explained in various ways. Surely with all the vast mountain country of the north there is room for the lemmings if they could only find it. But instead they go into the sea.

Peasants believe that they have a mysterious wish to commit suicide. Other people have suggested that the march to the sea is an instinct surviving from some ancient time when there was dry land where the Baltic and North Seas now are, and think the animals are aiming for this vanished land.

The beginning of the trek is almost certainly due to lack of food and to crowded conditions which their unsociable dispositions cannot tolerate. This accounts for the start; possibly the end is due to a blind effort to find some other home like the one they have left behind in the mountains, but which they never regain.

9

THE ANT-EATERS AND ARMADILLOS

UNTIL two Portuguese brothers landed in 1552 with "seven cows and one bull," Argentina had never seen cattle. Two hundred and fifty thousand square miles of the world's most fertile natural grazing grounds lay unused by the animals that to-day play such a prominent rôle in Argentine fortunes.

But with settlement came great changes; huge cities sprang up and the people of Europe and America heard tales of the South American "cattle kings" and "wheat kings."

To-day, South America presents a most varied and interesting picture. In the cities one finds stores, cafés, theaters, and every sort of gayety. There are people who live as like Parisians as possible, ordering their clothes and jewels from Paris; doing all the things that gay, rich people can think of to do with their money and time. Outside, in the country districts, one comes across huge estates, run like the estates of feudal times. In the fashionable places of Europe you may meet the owners, often titled, of these haciendas, and they will tell you of their life when they are at home; of how the lord of the estate holds absolute power over his employees; of his daily inspection, on horseback, of his vast acres and his direction of his farms and fields. While in the dwellings, the mistress of the house directs her maid servants in their tasks, and even sits with them while they weave the cloth used in their own clothes.

Farther away still from civilization and the cities are the wild places of South America; places that can only be reached by weary horseback or mule journeys, and that are still dangerous to the foreigner because of inimical natives, and impassable, disease-ridden country.

With the coming of colonists, the wild things at first came to the edges of the settlements to prey upon the herds; then gradually they made way for man and his gun. But South America is so vast that one animal still lives there quite unaffected by man and his ways, because he never encounters man. This is the ant-eater. His home is so wild that only the most intrepid and curious explorer can get at it. He does no harm to man and perhaps he does no good, although he does help to destroy a pest. But a pest so numerous that even his efforts seem unavailing.

The ant hills of North America are small, not as big as golf tees, and little processions of tiny black creatures go busily back and forth, in and out of their fragile sandy homes. But in the tropics of Central and South America there are nightmare ant hills, twelve feet high, built of hard, dry earth and divided inside into complicated long galleries and rooms. There are ant homes like small barrels, made of cemented wood fibers and hanging on the branches of trees or built inside their decaying trunks. And in each of these giant structures live thousands of the white creature known as the termite—a creature that is called the "white ant" although it is unrelated to the true ant.

The termites work stealthily, always under cover and often at night, for they dislike light. They feed chiefly on dead wood. Dead trees, dead lumber, house wood, furniture, all these provide a tasty meal for their little nibbling jaws. They work cleverly too, getting inside whatever they are about to destroy and doing their feeding there, leaving on the outside a paper-thin shell which gives the illusion of solidity and no hint whatever that within there is nothing but hole!

However, like most animal pests, the termites themselves provide a food supply for a larger creature—the ant-eater.

Not the banded ant-eater of Australia, which is a member of the kangaroo tribe, nor the spiny ant-eater of that country, which is the only relative of the duck-billed platypus. The enemies of the white ant are the real ant-eaters, of which there are none at all in Australia.

These true ant-eaters belong to an order of mammals called the toothless ones—Edentata. To this also belongs those armored creatures, the armadillos, even although the armadillos have peg-like teeth. And a third member of the group is the sluggish tree-inhabitant, the sloth.

Long ago, when the earliest man looked around him, he beheld the very ancient relatives of this order—the huge, lumbering, sloth-like Megatherium and the armored Glyptodon. But the creatures he saw were enormous, some of them eighteen feet high, reaching their heads up into the high branches of the trees to eat the leaves. Some of them walked around encased in a heavy, solid and inflexible armor. These are gone from our earth, but their relatives remain—the ant-eaters, armadillos and sloths—in tropical America. Those ancient animals were real giants, whereas the animal to-day known as the Giant Armadillo is only about a yard long, and the Giant Ant-eater is only seven feet!

Although truly toothless, the ant-eaters have a long,

narrow and sticky tongue which does much of the work ordinarily done by teeth. They can dart this tongue out of their mouths for many inches and upon it they gather insects, chiefly the swarming termite. The tongue is well supplied with gluey saliva which both attracts and traps the unfortunate insects who are later to enter the ant-eater's mouth as his next meal.

In the tropical swamps and forests lives the giant or great ant-eater. Other ant-eaters climb trees, but he lives entirely on the ground and eats no food but termites. With one rip of his powerful front claws he opens their rubbly nest and discloses a plenteous content of ants, larvæ and eggs.

In spite of his homely, very narrow head, and long tapering snout, he is a handsome animal. He stands over two feet high, with a coat of long, dense and drooping fur and thick bushy tail.

The strong claws of his forefeet act not only as a ripping tool, but also as a most effective and cruel weapon of defense against any animal that may attack him.

The silky, or two-toed, ant-eater is a climber, unlike his giant relative who lives on the ground. The two-toed animal gets his name from the two grasping and powerful toes of his forefeet, which, like the four toes of his hind, can close on the palm. To further aid him in climbing, the silky-haired tail is prehensile, like a monkey's tail. The animal is only about the size of a rat, but due to its long hair is much more pleasant to look upon than that unpopular object.

The Tamandua ant-eater of the tropical American forests has feet equipped for both walking and climbing, and is a strange combination as to method of progression. It walks only on the toes of its forefoot, but on the whole of its hind. Its tail too is prehensile, and covered, not only with short hair, but with small scales.

In Africa and tropical Asia there lives an interesting family once grouped by scientists with the Edentata, because its members have no teeth and because they eat insects. These are the pangolins. They too have weak

THE PANGOLIN

jaws and long, slender tongues, but unlike the ant-eaters, although like the armadillos, they are clad in armor. This armor is not like armadillo armor, which consists of horn-like plates. The pangolin is clad in a beautiful coat of mail. Hard scales, apparently composed of cemented hair, overlap each other like a symmetrical series of hardened leaves. Even the legs and the long, heavy tail are covered by this coat of mail. The creature in motion is a most peculiar sight. On the branches of the trees it moves about on its palms and soles, like other animals, but on the ground, its long front claws are an inconvenience, and it progresses on its knuckles! When one of these pangolins awkwardly ambles over the ground, its mailed tail dragging behind it, you almost expect to hear the clanking of heavy metal.

Also associated with the toothless mammals because

THE AARD VARK

he has a long extensile tongue and eats ants, is the aard vark, or African ant bear—a small creature with a pig's snout and long rabbit ears. But he is not at all like other ant-eaters. His jaws, although owning no front teeth, have back ones. He is heavy and about six feet long, coarsely furry, with strong legs and huge claws. The foot situation in the Tamandua ant-eater is reversed in the aard vark, whose forefeet are plantigrade, while he walks on the toes of the hind ones.

An indolent, slow-moving relative of the ant-eaters, almost idiotic in expression, is that unattractive tree animal, the sloth of South America.

Upside down, its claws dug into the branch, its head between its forelegs, its grayish fur looking like part of the tree, this animal will sleep for hours at a time, motionless as the branch it clings to. Although surprisingly enough it can swim quite easily, the creature is awkward on the ground and the widespread hind legs, dragging along after it, give it a particularly ungainly look. Its hair is untidy and its face vacant.

The sloths have voices; chiefly a mournful, senseless wail, and if injured they emit a whistle. No matter what they do or what happens to them, their faces remain as vapid and empty as ever. They live on the leaves and sprouts of the silvery imbuia tree. And, as if it recognized in them something so inactive as to be just about dead, on their fur there lives a small moth!

About another member of this tribe, the Maya Indians of Yucatan have a queer old legend. The vulture, they say, when it gets old comes down from the air and sits before a hole in the ground. It then sends out word to the other vultures that it is there, and they supply it with food. For a long time it sits there and gradually its feathers and its wings disappear. The old vulture has turned into an armadillo! When this transformation is complete, it goes into the hole and begins life as an armadillo. The Indians say this legend must be true because the vulture's bald head and the armadillo's bald head are so much alike!

The armadillo comes to North America too, and we who live in North America may have seen it in Texas. From its shell-like body-armor emerge a pointed snout and short, awkward legs. It jogs around like a small armored pig, its large claws digging its underground passages.

The animal is solitary, although sometimes one sees a small group of very young ones. Like the aard vark, it has no front teeth. It lives on worms and insects. It can run fast on its short legs, but not for any length of

THE ARMADILLO

time, and its main defense in case of attack is the hard shell which protects the softer under parts of the body, and the ability to burrow rapidly into the ground.

In South America we find that reckless, picturesque plainsman, the gaucho—the true native of the pampas. Like all natives, he learned early to utilize all the living things of his home, and among them was the armadillo. The live armadillo is hard to pull out of his hole, but the gaucho gets around this by sinking a tin cylinder into the ground and dropping in a piece of meat. Lured into the tin by the meat, the armadillo is caught, for the tin is too smooth for him to climb out. Once caught, the gaucho roasts or bakes him and lifts him from his shell like a hard-boiled egg.

Unlike many animals, the armadillos never change their clothes. Seasons come and go but the armadillo carries its heavy armor regardless of temperature. Nor does the appearance of the two sexes differ noticeably. In color they range from black to brown or yellow. There are several families, varying in size from the six-inch pichi to the yard-long giant armadillo. The armadillo armor, as is indicated by the fact that they can curl up, is not a solid mass, as was the armor of their ancient relative the Glyptodon, but some of the plates are jointed and flexible. The "ball armadillos" when frightened curl up so tightly that the softer parts of their bodies are entirely protected by their hard armor.

The giant armadillo is tremendously strong. It is worth your life to get this animal out of its burrow. It will fight to stay in and its powers of resistance are very fine indeed. On the short, strong legs, it can also make very fair speed when so inclined. Although classed as toothless, this animal has teeth; even the tongue has innumerable quantities of horny little points. The armor is composed of about thirty-six rows of square plates, eleven of the rows flexible. They are about the size of a thumb nail and in the center of each is a faintly outlined design. The whole arrangement is highly artificial-looking. The plates and pattern are one of those incredible examples of the natural symmetry and beauty of design so often seen in the animal kingdom.

This animal is a strange combination of the beautiful and the grotesque, with its handsome coat of armor, and its waddling, scuttling legs. Judging from the energy and power with which it uses its sharp strong claws to dig in the earth, and, like the ant-eater, to rip open termite hills, it would be a highly unpleasant enemy to encounter and hardly worth one's while to try to take anywhere it felt disinclined to go!

The familiar, active little armadillo of Texas and Mexico deserves our protection, for it eats worms and insects that are injurious to crops and gardens.

10

THE LION

When David Livingstone undertook his journey into the heart of Africa, in search of the sources of the Nile, he faced a thousand dangers and obstacles. In his day, exploration meant trudging on foot over unknown trails and through untracked jungle. Equipment was crude and limited. Little was known of the diseases that struck men down in the tropics, and remedies against the ills and fevers of the swamps were undiscovered. Livingstone met every handicap. He warded off illness; he avoided encounters with wild beasts; he had, no less, to protect himself against attack from the savages through whose country he traveled. They had never seen a white man before, and they were terrified of his light skin. Some thought him a spirit, appearing in the form of a man. He met their suspicions with friendliness, for while he had a will of iron, he was the gentlest of men.

News of Livingstone's progress was eagerly read. Then came a time when the news stopped. No word was heard from him for months on end. Interest in his fate was so keen that a searching party was organized, headed by Henry Stanley, and after two years Livingstone was heard from once more. Stanley had found him, an aging and broken man, quartered in a remote, native village, too feeble to manage the homeward journey alone. His rescue brought both him and Stanley to immortality.

Livingstone's courageous exploits in Africa won him the respect and admiration of the world. People read the fascinating story of his adventure with enthusiasm, but at one point in his book they stopped, in angry surprise. The great hero became the center of a storm of criticism. For Livingstone described the King of Beasts, the Monarch of the Plains, the African lion, as "a slinking coward, running off with his tail between his legs." He was violently contradicted—especially by those who had never seen a wild lion. The stay-at-homes knew the beast as he looks at the zoo, imposing and terrifying, his massive head framed in a heavy mane, pacing up and down the length of his cage with nervous energy, the picture of pent-up ferocity. In their minds, he was an heroic and awe-inspiring figure. He stood for dignity, for grandeur, and, above all else, for courage. They were shocked at Livingstone's insult to "the bravest and noblest of beasts," and the grudge against him on this score has endured.

But the hunters and old-timers of Africa know better. They have seen the lion on the veldt. They have watched him stealing like a shadow through the golden grass of the plains, lying sprawled on a sun-baked ledge in the stony kopjes, or crouching among the reeds at the edge of a marsh. They have seen his kill. They have seen him at play. Many of them have even faced him in the fearful moment of his charge. They realize he is a creature of changing moods. They neither hail him as a hero nor despise him as a coward. These men speak of lions very simply, using the name that best describes them—Big Cats.

Pick up your pet cat some day and look him over.

The lion has the same long, slender body, the same broadly arched skull, the same kind of teeth, the same soft pads soling his paws, and the same retractile claws. These are the distinguishing marks of cats the world over, big or little, wild or tame. They make them all

"BIG CATS"—BASKING IN THE SUN

kin, the tawny lion, the striped tiger, the spotted leopard and the forty other plain-coated or marbled and mottled wildcats, as well as the mild, fireside kitten. The lion is the only cat that grows a mane; the only one whose tail ends in a tuft of hair with a horny spike hidden in it. And of them all, he is the only one that roars.

Many other animals are equipped with claws, but only the cats have the power of drawing them in and thrusting them out at will. The lion, like all the other felines, walks softly on the pads of his feet, keeping his claws sheathed until he has need of their strength and sharpness.

The lion's name stands at the head of the cat tribe wherever it is listed, although his right to that honor is disputed by tiger enthusiasts. Many people contend that the tiger equals, and even excels, the lion in weight, in size and certainly in beauty. Champions of the two great cats will argue the points of their favorites endlessly. Their beauty is a matter of opinion. Their size is a matter of measurement, but both animals vary considerably. There are big lions and small lions, big tigers and small tigers. The question frequently arises as to which of the two cat kings would win in a fight. It is a question that cannot be decided, because in their normal existence these animals would never meet. If a lion and a tiger, brought together in captivity, were goaded to battle, the outcome would doubtless depend on which of the individuals happened to outweigh the other.

As subjects of conversation the two are rivals. In real life they are not. They live too far apart. Boys who hope to prove their courage by "going to Africa to hunt lions and tigers" will have to change their plans. There are no tigers in Africa. There is no evidence that they ever lived there, or anywhere outside of Asia in their present form.

The western world regards the lion as the embodiment of fighting courage, but in the Orient the tiger is the symbol for this quality. In India the huge cat of the gleaming golden coat with its broad, dark stripes is held in awe. Most people think of the Bengal tiger when the name is mentioned, but there is a tiger of the snows, too—the Siberian tiger—longer in build than his southern brother, with a denser coat of fur, shorter and more massive legs, and with his underbody quite white.

In Chinese legend the tiger stands as the model of supreme fearlessness. Mothers tell their children they must copy his bold spirit. Men going out to battle in the old days wore jade amulets carved in the form of the tiger to remind them of the strong heart that becomes a warrior. The title of kingship in the cat tribe is, therefore, only a matter of geography. In the main, the lion and the tiger are so much alike in their habits and build that an acquaintance with the one supplies a knowledge of the other. As a swimmer and tree-climber, the tiger far outdoes the lion, but otherwise they follow the same pursuits.

As far as tigers are concerned, lions have the whole field of Africa to themselves. The only lions outside of Africa are a few that live in a preserve in India in the Gir Forest, northwest of Bombay, a territory about four hundred square miles in extent, where they range within safe boundaries. It is forbidden ground to the hunter. In Persia, lions are no longer seen on the open table land, but up to fifty years ago they were reported along some of the remote river banks toward the south, where the Assyrian kings long ago found danger and diversion hunting them.

In Africa, each tribe has its own name for him, but all over the continent he is recognized under the beautiful name of "Simba." This is what the Swahili, natives of the east coast, call him. The early Arab traders, coming in distant days to barter for hides and ivory and slaves, opened their African commerce with the Swahili, and learned a smattering of their language as they bickered over their bargains. When the merchants journeyed inland, to extend their trading among the farther tribes, they took Swahili men with them as guides. Many of their words were picked up by the strange tribes with whom they came in contact, and in time the Swahili tongue became the "trading language" of the whole continent. There was every reason why the word "Simba" should become universally understood! Wherever the travelers went, their first question was whether they might encounter him on their way. His yellow form might be hidden in the grass. At any moment,

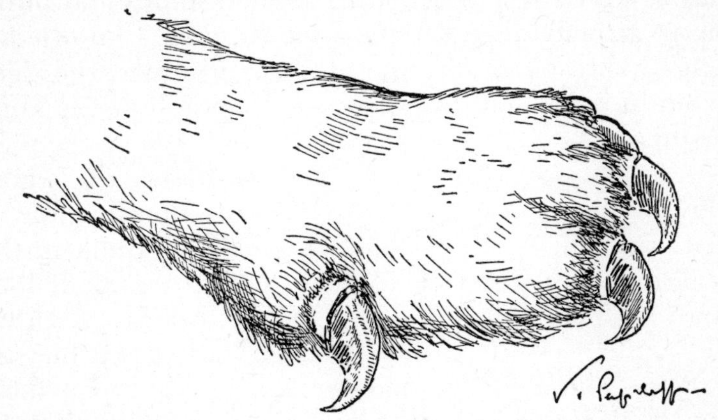

THE LION'S CLAWS

without warning and without sound, he might step out from the thickets. He might turn at their approach and disappear, and then, again, he might stand his ground. They never could measure their danger. At night, when they pitched camp, sentries kept a bright blaze burning to frighten him away, for lions have a dread of fire. By day and by night, they were always on the alert.

For in those times, the lion had his hunting ground over every inch of the African wilderness, except in the dry desert and the tropical rain forest. There the grazing herds on which he preyed did not exist, and there Simba was not found.

About two hundred and fifty years ago, the history of

savage Africa and all her wild creatures began to change. European settlers, equipped with firearms, entered the southern area and transformed it into a stretch of farms and small towns. They had to have peace and safety. They had to have food. With their guns, they drove the herds of zebra and antelope that crowded the veldt back from the coast region. Where the herds went, the lion went, following his food supply. The kingdom of which he was monarch shrank. Lions no longer roamed at will from one end of the continent to the other. From year to year the double retreat of wild game and wild lions continues. To-day lion country is limited to the central stretch of wilderness between Algeria and Bechuanaland and the unsettled districts of East Africa.

Many people imagine Africa as a monotonous clutter of dark, tangled jungle. It is really a very beautiful country, with every variety of scene and climate. There are soft, rolling hills. There are high plateaus, where the nights are frosty. There is the bush country, bordered with thickets. There is the burning veldt, where the pitiless sun beats down on the bare and dry plains. In all of these regions lions may live.

It used to be thought that the nature of his habitat had some influence on his appearance, especially on his mane—that those of the high plateaus developed thick manes as a protection against the cold, and that those of the thickets carried scanter growths as a result of the tearing of the brambles and the thorn bush, but it is now recognized that manes are a matter of accident rather than locality. The biggest of wild lions may have the scantest of manes. Even cubs of the same mother may have manes of varying size, and of varying color. Rarest and handsomest of all are the "black-maned" lions. They are not a separate variety, however, and one of them may be brother to a tow-head.

It is the dream of every hunter who goes out after lions that he will be lucky and bag one with a "good" mane. The chances are that he will never fully realize his dream, for no matter how rich and thick a wild lion's mane is, it almost never measures up to the grand, shaggy shock of hair that adorns the head and chest of the cap-

THE LION'S PADDED PAW

tive lion. It used to be held that the chief difference between the Indian and the African lion was that the former was virtually maneless, but lions from both countries reared in captivity develop equally heavy manes. No one knows for certain why zoo life has this effect, but it is probably due to the combined colder climate and dampness of the northern countries.

Life behind the bars works other changes. Restraint and rich feeding, with none of the effort of hunting, from cubhood on, deprives a lion of his real beauty. The Big Cat of the zoo is too round. He is clumsy, padded with fat. His muscles sag. Even his backbone changes, and takes on an arched curve, whereas the back of a wild lion has a dipping sweep. It is perhaps natural that they should differ in these respects, but no one can explain the startling contrast in the color of their eyes.

The zoo lion's eyes are a dark, slumbrous brown. Those of the wild lion are brilliant yellow, and shot with light.

Times have changed since the early Arab traders broke their cautious trails through Simba's domain. Nowadays when men go into his territory, they make most of the journey in automobiles, driving over the level plains where there are no roads. A trip into lion country is less dramatic than it used to be, but more interesting, because it affords glimpses of the Big Cats as they live from day to day, from hour to hour, in their own world. Many a lion has been under the gaze of man without suspecting it, thanks to the clear lens of long-distance glasses. Many a lion has turned his head and unknowingly looked straight into a far-off camera. Unaware of the presence of intruders, undisturbed, a troop of lions is a breath-taking sight—not because they are so fierce, but because they are so mild!

Simba in the daytime is the picture of peace. The hours of hot sunlight are his hours of ease and play and rest. He takes his deep sleep in the early morning, in the cool dark shelter of the reeds, in the cleft of a rock, or in some thicket retreat, coming out into the open to doze and drowse through all the rest of the day. He is tired after the strenuous hunt of the night before; he is full-fed; he has quenched his thirst before dawn. What else is there to do in the burning heat of an African day but sleep? Sometimes he wakes up, turns his head lazily, opens his great jaws in a gaping yawn, and drops back into his doze. It is too hot to move. The lioness keeps an eye on her cubs, as they roll and wrestle and spar in a mild boxing match. She plays with them, deals them little pats of affection with her heavy paws, and cuffs them in punishment when they get too rough. Big

Simba looks on, too lazy to take any part in the frolic. Sometimes the family circle includes friends, all on pleasant terms.

SIMBA AT NIGHT

Now and then one of the group may rise to stretch his legs in a short promenade, or the whole party moves languidly toward the inviting shade of the trees. The far-off camera can now settle the ancient dispute as to whether lions can climb trees. They were formerly supposed to be quite lacking in this cat accomplishment, but very often some young lightweight will claw his way up to stretch full-length on a low-hanging bough. The big chaps keep to the ground, no doubt because they are too heavy for easy climbing.

Even his victims of the night know that Simba need not be feared in his mild daytime mood. Time after time, men have seen a lion walk across the plains in the full glare of the sun, straight toward a herd of grazing zebra and antelope who saw him in full view, but gave no signs of alarm! As the lion came nearer, some of the animals stopped their grazing and moved quietly to one side, leaving a path in their ranks for his passage. Some of them went right on with their cropping, scarcely raising their heads. The lion walked by as though the plains were empty.

Of course, this doesn't mean that the lion's widespread reputation for ferocity is unfounded. He can be the most deadly of killers, the most formidable of fighters, when he is aroused. His strength is prodigious. The power of his bite, the clutch of his claws, the force of his spring are weapons that terrorize both man and beast. He is to be dreaded when he is fierce, dreaded when he is mild. There is always the chance that he may give battle. There are no rules for lion behavior. You never can tell what a savage lion will do. He knows his own world, and is master of it. But when man invades it, he grows uneasy. Man is something from the outside. He makes strange sounds; he brings pain; he sets the grass of the plains afire, spreading suffocating flames. He interrupts noonday naps with the sharp reports of his rifle. Even his tiptoe tread is a signal of alarm, though usually the lion does not wait to hear it. Somehow, he gets his warning from afar.

The keenness of the lion's power of scent is much debated. Some hunters say it is not extremely acute. On the other hand, men of long experience in Africa insist that a lion can even distinguish between the scent of a white hunter and a native. They say that he is always more nervous at the approach of the white man. Africans themselves are creatures of the wild; they move almost as silently as animals. They wear no clothes; their feet have never known shoes; their tread is soft and stealthy. There is something of Africa itself about them. It is true that their scent is different from ours, and it is quite possible that the lion perceives the difference. But whether the man is white or black, the lion's instinct is to get away from him. Simba is no fool. He meets danger gallantly and boldly when he must, but he doesn't court it.

Even a good hunter may spend weeks in lion country before he catches a glimpse of one of the Big Cats. Many a lion has made for cover and vanished without ever being seen. In distant retreat, he is not only cautious; he is clever. He moves in a long, low-swung stride. His powerful muscles are a match for his heavy weight. A lion is four feet high at the shoulder, between nine and ten feet long from the soft, whitish beard on his chin to the black tuft on the tip of his tail, yet when he is on the move, he can make himself all but invisible. His yellow form and the yellow grass merge in a dim blur. Head low, body lithe, he eases over the rise and fall of the ground so evenly that nothing but a slight brushing of the grass betrays him.

Surprise encounters often occur. During the day, lions go about in troops, and a party of them may suddenly step out of a thicket or from behind a rock without the slightest sound. It is hard to say which is the more terrified—hunter or lion. What happens depends very much on the nature of the particular man and the particular lion. No two men are alike, and no two lions are alike, but hunters are trained to danger, and wild lions are trained to nothing at all. The whole troop may walk off. Any one of them may suddenly turn and come back.

The wisest thing for a hunter to do who finds himself staring into the yellow eyes of a wild lion a few yards away is to stand perfectly still. He dare not raise his gun. There is no way of telling whether he is face to face with an irritable lion or a phlegmatic one; whether the beast will quietly make for cover, or charge. The hunter can only hope for luck, and pray that his gun-bearer or companion (for no one ever goes into lion country alone) will shoot straight, and shoot to kill. It is when a lion is wounded that he is most dangerous. The pain multiplies his strength, raises his vitality to the highest pitch. At the moment when men fear him most, they admire him most. He is a die-hard. He fights to the last gasp. Even in the agony of a mortal wound, the grand beast summons up his last ounce of might for one final spring, and it has happened that at the point of death a lion has leaped and brought his enemy down. More than one hunter has gone up to inspect his trophy, thinking his lion dead, and been killed by a last stroke.

The lioness can be even fiercer than the male. If her mate is attacked, she takes up the challenge with deadly fury. Hunters, coming on a pair, always try to bring the lioness down first. She is the more quickly roused and the more dangerous, lighter in weight, and swifter in the charge.

Lions do their own hunting at night, depending on a stealthy approach and a sudden spring rather than on open pursuit. The zebra and the antelope are their favorite food, though a hungry lion takes what he is lucky enough to find. The vast game herds of Africa have been heavily diminished, and the night's work is hard and often disappointing. A lion may have to travel fifteen miles before he hits the trail of his victims. His best chance is at a waterhole, where the animals gather after sundown to drink. They are restless and wary, alive to every rustle, and shift about uneasily for a long time before they venture to lower their heads. The lion's leap is only about fifteen feet, and he must bide his time until the perfect moment, for at the merest suspicion of a sound, the group at the pool will scatter in panic. The impact of his weight and the blow of his powerful paws as he lands on his victim have a numbing effect, but the actual killing is done by the bite of his teeth. He seems to know the vulnerable spot, for he generally places his bite accurately, breaking the vertebræ at the base of the neck.

Very, very few men have actually witnessed the lion's kill, but much has been discovered by examining the remainder of the carcass on which a Big Cat has fed. Also, observers have taken up a post of watch in *bomas,* or blinds, constructed of a mass of thick bush and thorn trees, and left a dead zebra near by as bait, so that they might study lion habits. People used to believe that a wild lion would not touch meat unless he himself had brought it down, but he is quite content to feast on any meat that he is lucky enough to stumble on in his prowls, even if it is stale.

It is Simba's habit to take his meals in privacy. He drags the carcass off to some secluded spot in the bush, where there is little danger of intrusion. There he proceeds according to lion table-manners. He first removes any part of the entrails that offends his sense of smell, and often goes to the trouble of covering up the discarded parts with earth and leaves. He does not bother to chew his food, but tears off the meat in great chunks and gulps them down whole. His tongue is covered toward the back with rough papillæ, which scrape the bones clean. A lick at a man's hand would tear the flesh deeply.

One thing no lion likes when he is feeding, and that is uninvited guests at his banquet. His mate and cubs may join him, if the lioness has not made her own kill, but if another lion should attempt to take pot luck with him, there is sure to be a protest. The intruder is usually driven off by a threatening growl, but if he fails to take the hint that he is unwelcome, he finds himself involved in a battle. If several lions have combined to bring down a large victim, they share in the spoils. Even then, quarrels may occur, but lions rarely kill each other.

The sneaking hyena, never a hunter on his own account, is always on the watch for a free meal. Somehow, he gets the news of the kill, and hovers around until the lion has eaten his fill. Three-quarters of a zebra is as much as a lion wants, unless he has had a night or two of starvation. The hyena and the swarming vultures feast on the leftovers. If a hyena's hunger should get the better of his fear, he may presume to approach while the lion is still feeding. Then he may not live to regret his boldness, for the lion will interrupt his meal to put an end to him. Otherwise, he does not kill. Leave him at peace, and he keeps the peace.

After the great beast's hunger has been satisfied, he must drink, and it is then, on his way to water, that he lets forth his roar. He begins with a succession of short growls, abrupt coughing jerks of sound, which grow louder and more prolonged, swelling into reverberating waves of tone. Hearing the roar of one lion, others answer, and their booming, like an orchestra of drums, shatters the stillness of the night a mile away. As they carry their heads low when walking, the echo of the roar travels over the crust of the earth, gathering volume till it matches the vibrating of thunder and the pounding surf of the sea. They are noisiest on cloudy, rainy nights. The hunter in his camp, the planter in his cabin, the native in his hut are awakened by the distant rumbling. Old-timers are not frightened by it, because they know it is not a threat of Simba's approach. When he comes on his stealthy errands, he is silent. A lion may roar when he is balked of his prey, but not when he is tracking it. The blood-curdling sound carries no menace. It may be a call to his mate, a signal to his friends, or merely an expression of well-being after a good dinner. As one hunter has put it, "The lion often roars for much the same reason that a cowboy yodels—out of high spirits."

There are a great many stories about the lion which are firmly believed, though they are not borne out by the facts. It is pleasant to picture Simba as faring forth to win his mate in combat, leading his chosen one off to his lair, and living happily ever after, devoted to her alone to the end of their days. There is only one thing wrong with this romantic description of the Big Cat's married life—it isn't true. Lions are polygamous. It is true that they often fight over a lioness, and that a pair often remain together in the same troop for a while. The mates frequently continue their companionship during the infancy of their cubs. But a troop may consist of one lion and several females, of a number of males and a lioness or two, or two or three lions may wander about without the company of females at all. Any sort of combination is possible.

Another current tale concerns the "man-eater." It is often said that once a lion has tasted the flesh of man, he is content with none other. It is true that man-eating lions have terrorized villages, and that they are astonishingly daring in their attacks. But "man-eaters" do not come in search of human prey because of a preference for it. They are almost always old lions, whose strength is not equal to the hunt for swifter game. On plantations and farms, they sometimes attack the cattle, which in their corrals are surer victims. Or they steal up to a camp, attracted by the smell of meat in the kitchens or of the hunter's kill. Hunger makes them desperate, and they spring upon the first meat their nostrils track, man or beast.

The life of an aging lion is none too pleasant. He loses his power and skill. His muscles grow slack. His spring is shortened. His eyes weaken; his teeth grow soft; his claws become dulled and feeble. He is worn out with a lifetime of hunting, night after night, and with his occasional battles against his rivals. He begins to look for a share in the kills of other lions. The younger males resent his presence, and drive him out to fend for himself.

All the privileges are for the cubs. The lioness looks after them, nursing them when they are tiny, teaching them to feed on the kill she provides, and later instructing them in the art of hunting. A lioness with cubs is always on the alert for their safety. She is very likely to attack without provocation, simply because of her instinct to protect them. By the time the cub is six years old, he has grown into full lionhood. The spots on his coat have disappeared, leaving him an even tan, though once in a while this mottling of babyhood only fades, leaving giant freckles to darken his sides for the rest of his life. He seeks out his own friends, joins another troop, fights the older lions for the company of the lioness—he is Simba, the stalking hunter, when he is hungry, the Big Cat basking in the grass when the sun is high. He kills for his food, for his life, but not otherwise.

Among African tribesmen, the lion is a symbol of all the qualities that a man should have—fighting strength and a fighting heart. Chieftains call their eldest sons after him, and sometimes a full-grown man who has proved his valor will assume the title "Simba" as his right, discarding his own name. Nothing so distinguishes a hero as a triumphant conquest of a lion. Among the Masai, who are a valiant tribe, the wearing of a headdress made of a lion's mane is the highest badge of honor and when a youth has won such a trophy, he is admitted to the ranks of the "warrior class."

Before the dawn, a band of naked youths, armed only with shields of hide and spears of iron, set out to win their glory. They have heard Simba's roar during the night, and know the direction of his prowls, but long before their coming he has withdrawn to cover. Cautious and noiseless as Simba himself, the hunters steal on their way, watching every clump of thicket, every spot of brush for a betraying sign of their quarry. The lion is at this hour disinclined to fight. He is worn with the night's hunting, heavy with his feeding. The Big Cat wants sleep, not action. The hunters want their headdress! They seek him out, stealing around him in a close circle, their shields before them, their spears poised to meet a sudden charge. They challenge him with shouts and taunts, leaping back and forth, trying to confuse him, so that he will not know which way to spring. He is bewildered by the yelling and the moving bodies.

His topaz eyes shift from man to man. The shouting grows wilder and more tumultuous. Simba is uncertain. He draws back, growling in quick, deep jerks. He stands at bay. Suddenly, his muscles tense. His mane bristles. He seems to swell in size. He lashes his tail from side to side, in strong, nervous strokes. In an instant, it stiffens, and the Masai know that he is ready to spring. They never take their eyes from the beast. Their hands tighten on their spears. As the lion leaps, the man facing his attack thrusts his spear with all his might, and drops to the ground, covering his body with his shield, so that the hurtling beast may miss him as he lands. The instant the brave youth falls, his spear quivering in Simba's side, all the others press forward and hurl their weapons full force. It is a terrifying moment for even the stoutest of fighting hearts, but the Masai never fail their leader. The lion's ferocity rises highest when he is wounded, and the Masais' courage soars to meet it. They know Simba's inflamed strength, his tearing claws, his deep-biting, poisoning teeth. Closer and closer they press, repeating their spear thrusts. The lion fights to the last—but in the end, he falls.

The man who first hurled his spear gets one half of the dead lion's mane. The other goes to the youth who first managed to grasp Simba by the tail. Proudly, the warriors flaunt their headdresses at their feasts and dances, where the story of the combat is told and retold.

The Nandi spearmen celebrate their lion hunts impressively. When they have brought one of the great beasts low, they form a circle around his fallen body and with their shields raised aloft, so that their shadow dims the once-dreaded form, they sound a chant in Simba's praise. Their song is a mixture of respect for his courage and jubilant triumph over his defeat. To them, he is a foe worthy of their spears, a gallant fighter, brave against all odds, unyielding to the last, though he did not seek the battle.

Was Livingstone right when he branded the lion a slinking coward?

11

THE CHAMOIS

One day in the fifteenth century, the emperor of a mountain kingdom went out from his fortress-like castle, perched high on a Tyrolean cliff, to hunt the chamois.

It was very different from the modern hunt. He did not carry a rifle, nor was he dressed in the comfortable tweed jacket and soft leather shorts of the modern chamois stalker. Instead, there issued from the castle gates a long procession. First came the emperor, dressed in the kilted skirts and full-sleeved jerkin of those days. Then the gentlemen of his court, and the stalwart men-at-arms who were to do the hard work. After them, trailed the ladies of the court, dressed not at all to scale lofty peaks, but in long silk gowns and flowing head-dresses. Accompanying this parade, there was usually a chaplain who went along as protection against the evil spirits of the mountain. These evil spirits were mysterious beings whose spells lured the hunter to the edge of abysses and over. They caused avalanches, stone slides and thunderstorms. Or they appeared, ghost-like, at midnight among the moonlit pinnacles of rock near the highest mountain passes.

In these same Tyrolean Alps to-day there are people who half believe in tales of mountain magic. And when there is a thunderstorm, you can hear the church bells being rung in supplication for protection for the people and the crops. Surely this is an inheritance from the days when such supplications were made to those sorcerers who were held responsible for all the accidents among the mountains.

In the far-off days when the mountains were less explored, a great many people indeed credited all the current legends. And at any rate, whether the kings and princes believed in magic or not, they took along a chaplain who could exorcise the evil spirits, "just in case."

As a matter of fact, many of the falls and injuries the chamois hunter suffered were merely due to the ability of the animal to make its way securely about the narrowest and most precipitous of ledges. This often led over-enthusiastic hunters, as it once led this very emperor, into positions of great peril. Maximilian tells of one adventure when he pursued a chamois with too much eagerness, and losing track of time, found darkness descending upon him. At the same moment, he realized that he was on a narrow ledge of rock above a black abyss. Such a narrow ledge that he could neither go on nor turn back. The chamois had gone blithely skipping around the corner of the rock on its small secure feet, but the man stood there afraid to move lest he should lose his balance and hurtle down into the darkness beneath him. He does not tell us what finally happened. But some one must have found and rescued him, for he lived to write about the adventure.

But from two evils of the mountains, the chamois itself was not safe. These were the two huge birds of prey, now almost extinct,—the Golden Eagle, and that vulture-like bird, the Laemmergeier. More frequent than the oft-told tale of the eagle who carried away babies while their parents worked in the fields, were the eye-witness accounts of the Laemmergeier bearing in his talons the soft, helpless body of a young chamois.

However, on such a hunt as this, the emperor had no idea of scaling dangerous peaks. Up the jagged rocks, high above timber line, grazed the herd of delicately built, fawn-colored animals. Up the same rocks, slipping and stumbling in spite of the "crampons" whose heavy, six-spiked soles gave them a foothold, labored the emperor's beaters. When they arrived near the herd, they circled and drew in upon the animals until little by little the chamois were forced down. Farther and farther down, until at last the terrified little animals found themselves on the shores of a mountain lake!

On these shores was seated the audience—the emperor and his nobles and ladies. At this point in the drive, the gentlemen of the party arose and stepped into boats in which they were rowed out upon the lake. Servants then drove the bewildered chamois into the water. And from the boats the court gentlemen threw javelins at the chamois until all the animals were killed. This was a fine day's amusement!

Such a hunt was more or less of a picnic. The same emperor, Maximilian, called the Last of the Knights, was a bold hunter. His knowledge of this sport has come down to us in a book he wrote. It shows that he was a man of considerable common sense who went in for sport in what in those days was a very sportsmanlike way. Chamois hunting has always been dangerous. It was conducted then, as now, in two ways, either by drives or by stalking. A drive consists in having beaters circle in upon the animals and drive them past the hunters. Stalking is the more sporting method of the two—hunting the chamois unaided, except perhaps for a guide.

The danger lies in the difficulty of following the fleet-footed animal. Unlike many of the favorite game animals, the chamois is not startling in appearance. It is about the size of a large goat and looks very much like an antelope. The fore part of its body and its neck and head are slender; the hind part a trifle heavier. Its face is gentle and timid. The most easily recognizable character of the animal is its short, black and backwardly-curved horns. Both male and female have these horns. The legs are slender and the hoofed feet small. The animal apparently has a marvelous natural balance, and moves with a swiftness and security that baffle the most surefooted hunter.

The chief difference between fifteenth century hunting and that of more modern times is in the weapons used. The modern rifle was then unknown. The game of those days was hunted with crossbow, spear and javelin. One of Emperor Maximilian's crossbows is on exhibit to-day in a museum in Vienna. The steel bow is nearly four inches wide; the bow string is twenty-six inches long. To bend that bow it was necessary to use a portable hand windlass. In order to save time, these cumbersome weapons were carried already bent. The arrows have massive iron points. These were used for big game. For small game lead bullets were customary. These bullets were used, not in a gun, but in the crossbow, many of which were so constructed that they could use either arrow or bullet.

The javelin was a heavy weapon about eight feet long, and a strong skillful hunter could throw one of them twenty-five yards or so.

It is easy for the chamois to balance its slender body with the four small-hoofed feet on a tiny pinnacle of rock. But for a man to stand poised on a ledge and to aim and hurl a heavy javelin is not so easy. Especially as the chamois do not stay in position and let themselves be used as a target.

Like most wild animals, their sense of hearing and of sight are very acute. Their movements are remarkably swift. They can move around mountain peaks as easily as if they were on flat ground, and only one other mountain animal, the ibex, goes higher.

The chamois can make a sudden descending leap of twenty feet, keeping its balance by striking its back hoofs once or twice against the wall of rock during the descent, and alighting securely on a pointed bit of icy rock. The hind feet land first and the forefeet are brought back almost to meet them. As their feet are very small, this position takes up little room and enables the animal to stand safely where nothing else could get a foothold.

In the winter months, the chamois stay around the timber line and lower, but in the warm weather they go up into the snow of the great heights and when this time comes they put on a heavy winter coat. We know the chamois most intimately from the commercially prepared skin used to polish automobiles and ladies' faces. This is always a soft yellow. But we do not know the real color of the chamois at all unless we have seen one somewhere.

In the Swiss and Tyrolean Alps, the animal's summer coat of close fur is chestnut brown; in winter this coat is replaced by a heavier one, lighter and much grayer in color. The chamois of the Pyrenees, the chamois of the Caucasus, and the chamois of the Carpathians and of the Alps all vary slightly each from the other in color. But all of them have black and white face markings and a black tail. Occasionally there are also variations among animals living in the same region. Very black chamois are rare and considered a great prize. White chamois also crop up sometimes, and for centuries it has been considered bad luck to kill one of these albino beasts.

The chamois herd is well organized. There are usually about fifteen or twenty does, young, yearlings and two-year-olds. The old males, except in the breeding season, are rather solitary animals. The animals live on herbs or pine shoots, and do most of their feeding in the early morning and in the evening. They are extremely fond of salt, and in some parts of the mountains where there is sandstone formation with an impregnation of salt, you will find rocks that have been deeply hollowed out by the licking tongue of the chamois.

When the herd is grazing, there is always a sentinel chamois. He usually stands slightly above the others and keeps a lookout for danger. At the slightest hint of an intruder, this sentinel gives a shrill, windy whistle, and in a second the herd has scattered in alarm! This whistle, to the veteran stalker who is able to conceal his whereabouts, is often a give-away of the presence of chamois.

Naturally, hunters prefer to get an old male, for they are harder to stalk because they move around alone. Also because they provide the hunter with a cherished trophy of the chamois hunt,—the "Gemsbart." This tuft of stiff hairs looks something like a shaving brush, and it is a most coveted ornament on a mountaineer's hat, for it means that the wearer has bagged an adult male chamois. It unfortunately also appears very often on the hats of other people as well—tourists who have never been farther up a mountain than the nearest funicular railway would carry them, and who have invested in a Gemsbart, or an imitation Gemsbart, in some shop in Interlaken, Bolzano, or Innsbruck. The word Gemsbart means chamois beard, from the German words *Gemse* (chamois) and *Bart* (beard). This would indicate that the blackish, white-tipped hairs came from beneath the chin of the chamois. But this is not the case. The chamois has no beard. These hairs come from along the ridge of the backbone of the old male when he is in his winter coat. And as the old male is wary and difficult

SENTINEL CHAMOIS

to get, an Alpine hunter considers this trophy well worth the trouble it takes to laboriously pull these hairs one by one from the animal's back after it has been brought down.

There was an epoch when chamois hunting descended to an incredible state. This was the century following the time of Maximilian. It was about this time that a great duke at festivities in honor of his marriage, sent out beaters to drive a herd of chamois on to cliffs from which even they could not retreat. The duke then had them shot at from below with howitzers!

At this time and during the following two centuries, hunting was still almost exclusively the sport of kings and nobles, to whom most of the hunting territory belonged. There was probably plenty of poaching and the penalties for this were extremely severe. Due to the difficulty in these mountainous regions of telling where one duke's land ended and another duke's land began, there were many bloody encounters between the hunters, servants, beaters and even guests of the landed proprietors.

An old copperplate, still in existence, shows one of these noblemen on a chamois hunt in the seventeenth century. This throws a good deal of light on the lazy way in which hunting was then carried on. The picture portrays a very fat nobleman, roped, climber-fashion, to two retainers. The two retainers stand on the top of a small hummock, perhaps intended by the artist to represent a mountain, tugging for all they are worth on the rope. Apparently they are making a valiant effort to haul their noble patron to the top of the hummock, which seems to be only some three or four feet above him. He, in his turn, is pushing intensely with an alpine stock. Whether he expects to vault to the top of the hummock, or to wedge himself more firmly in the safe ground beneath is left to our imagination.

In the nineteenth century, among royal hunters there arose one more like our modern conception of a sportsman. This was the Archduke John of Austria. He knew the mountains as a seasoned mountaineer knows them, and hunted with some intelligence, not for the mere adventure of killing some animal. He explored the peaks without a retinue or a company of beaters, and he gathered his knowledge from the sturdy peasants whom he met and made friends with on his climbs. It is interesting that this archduke eventually married the daughter of the keeper of a posting station on one of the passes. And although it cost him his position at court for a while, they "lived happily ever after."

To indulge in real chamois stalking with any pleasure or safety a man must have this thorough knowledge of

AN OLD BUCK

the mountains and of climbing, as well as a trained and hardened body. For such hunting, not so terribly long ago in the 1800's, a little half-English, half-Austrian boy decided to train himself. This boy lived in the Tyrolean Alps in one of the old castles. When he was quite small his mother feared that he was going to be a bookworm, and used to warn his teachers that he must get more exercise. But all the time, the boy himself was assiduously practicing with an ancient rifle, taller than himself; climbing around the lower peaks, and absorbing from the peasants of the neighborhood much useful knowledge and advice about climbing. William Baillie-Grohmann knew almost all of the landed nobles of the Austrian Tyrol, and hunted on their preserves, but he was also the friend of peasants and mountaineers. Due to the fact that he grew up speaking two languages, as well as the local mountain dialect, he was able to play the part of a seasoned Tyrolean mountaineer. With his bronzed skin and his air of being completely at home among the mountains, he had many amusing adventures with tourists. They often tried to tip him for giving them information. They even asked him to carry their luggage, and were very much embarrassed afterward to find that he not only was not a guide, and not a simple peasant, but that he spoke English as well as they did. Mr. Baillie-Grohmann is typical of the independent, skilled and intelligent sportsman of to-day.

Despite his passion for hunting, he believed in wise hunting, and his great grief was that such rapid depletion was going on in the chamois ranks. This was particularly true in Switzerland where for a time shooting was entirely open to any one who had a gun. Now, however, this over-shooting has been stopped, and measures have been taken everywhere to protect the chamois from extermination. In the very mountains of Baillie-Grohmann's home, there has recently come into being a preserve for chamois and ibex.

There are comparatively few royal families now, and few nobles can afford big hunting preserves. But one of the remaining kings, the King of Italy, in 1923 gave to the Italian Government an enormous game preserve which had been part of his personal property. This is now called the Gran Paradiso. It lies in the Italian Alps, which, of course, now include part of the former Austrian Alps. It is not a formal park, but has been patterned after our own national parks. Within this cordon, drawn around miles and miles of the most beautiful snow mountain region of the Tyrol, the chamois live protected among their own rocky haunts.

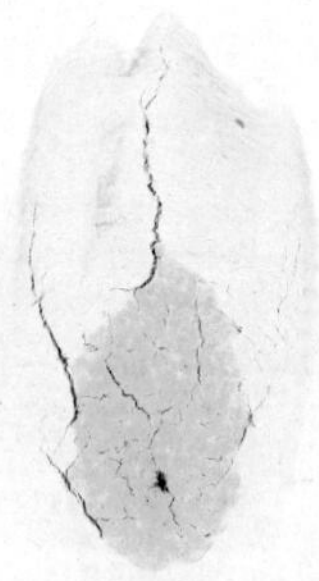

12

THE RHINOCEROS

On a brisk October day, nearly two hundred and fifty years ago, Sir John Evelyn, one of the leading gentlemen of London, sallied forth from his house to inspect "the unicorn," an animal imported from India by a group of merchants. All the fashionable folk of the city found time to go and stare at it. They discussed the beast from nose to tail over their gaming tables and at their barge parties and evening routs, for to Englishmen familiar with the grace of the deer and the elegance of the horse, he was a surprising combination of ugly and interesting features. Sir John was a friend of King Charles II, and though he disapproved of his monarch's frivolity and gay pleasures, they had much in common, for Charles had another side to his nature. He had a love of music and painting and a keen interest in science, and in his palace laboratory he copied the experiments of the day. He and the gentlemen of his court were constantly receiving curiosities from far-off lands, and "the unicorn" matched anything they had ever seen in strangeness.

In legend and fable, the unicorn is a make-believe animal, with the body of a horse, the wings of a monstrous bird and a single horn sweeping in a slender curve from the top of his head. But the creature that Sir John and his contemporaries gathered to see was not the imaginary beast, miraculously come to life. It was a rhinoceros, and they borrowed the name "unicorn" for it because it was an Indian rhinoceros and, therefore, single-horned.

The animal's more ancient name was supplied by the Greeks. The first thing that struck them when they saw the creature was, not his enormous barrel-like body, not his long cask-like head, not his thick, gray, hairless hide, but the very strange location of his horn. They were acquainted with horned animals, but all those they knew carried their natural weapons in graceful symmetry on either side of their brows. But this beast bore his upright, jutting out on the top of his nose! Let him be known by the one feature that distinguishes him from all other beasts that walk, they said. *Rhino* means nose, and *keras* means horn. They joined the two, and formed the word rhinoceros (as we spell it), christening him with the apt name, "nose-horn." When we speak of the brute as "the rhino," we are really giving him the undignified title of "the nose."

There are five distinct kinds of rhinoceros, each readily recognizable, three Asiatic, and two African. The Indian is the giant of the Asiatic trio. Next to the elephant, he is the largest of all land animals. Though he holds second place for size, it is commonly agreed that he should rank first in ugliness. He is without a single saving grace. In addition to the clumsiness of his two-and-a-half-ton body and his long, one-horned head, he has a uniquely hideous skin. He looks as though a rash of lumpy warts had broken out all over his gray person, but these "tubercles" are a normal part of his make-up. His skin hardly seems to fit his body, for it resembles a patch-work of badly joined slabs. The Indian rhino looks as though he were covered with thick plates of stippled plaster, spread in triangular areas over the shoulders and in squarish shields over his thighs, overlapping at the edges in rough folds. Around his neck is a collar of bulging rolls of skin.

Old-fashioned writers used to compare the Indian rhino's hide with a suit of armor, such as was worn by soldiers in mediæval times. From this comparison, the

INDIAN RHINOCEROS

idea got about that the animal was actually covered with a coat of mail. In Shakespeare's day a book was published with such a description of the rhinoceros. Every one took it for the truth, and very likely the author, too, believed it. How any animal could come by a metal coat, no one bothered to inquire. The book contained a picture of the "armor-plated" beast, showing him clad in a series of shields, studded with round rivets. These were supposed to hold the remarkable covering in place. The portrait was the work of one of the master artists of the world, Albrecht Dürer, who had undoubtedly intended it as a fanciful picture, with the rivets representing the lumpy tubercles on the skin. It was, however, generally regarded as a true-to-life portrait, and as it was reprinted time after time in natural history books, it advertised the rhino as a mail-covered animal more and more widely.

Dürer is supposed to have drawn his misleading rhino portrait from a sketch made of a real rhino, an individual with something of a history. In 1513, the King of Portugal shipped an Indian rhinoceros, which he had received as a present, to the Pope, thinking it would make a welcome gift. The animal, having sailed all the way across the Indian Ocean, should have been well used to the ups and downs of sea travel, but his second voyage seems to have upset him completely. The story is that while on shipboard he fell into a wild fit of temper, and rampaged around until he sank the boat and himself with it.

The idea that a rhino's hide is impenetrable still persists, although it is no more so than any other inch-thick hide. It can be easily cut with a pen-knife and a little muscle before it stiffens. A shot that strikes the heavy bone of the head may glance off, but the hide itself has no peculiar resisting power. A British soldier, stationed in India, once got himself into serious trouble by indulging in a bit of target practice aimed at a young rhino, which was kept in camp as the regiment mascot. He really thought his bullets would not penetrate, but the first one brought the rhino down, and the Tommy found himself in the guardhouse on a charge of misconduct.

Rhinos are a vanishing race, and from year to year their numbers grow fewer. Their destruction began as a necessity, when the wild land was converted into settled country, but it was not only because he was a nuisance and a menace that the Indian rhino was removed so liberally from the scene, but because his horn came to have a high money value.

It happens that rhinoceros horn is very different in texture and structure from the horn of any other animal. It has no relation to the hollow sheaths we know as "true horn," which is peculiar to the ox and his kin. It is totally unlike the antlers of the deer, and in no way resembles the skin-covered, bony "horn" of the giraffe. It is neither scale nor bone, but a hard mass of fused, agglutinated hairs, thickly grown together and cemented by a special cell formation, so that it forms a solid bulk. The nasal bones of the rhino skull are massively built, and support the weight of the horn firmly. The horn itself is not attached to the skull, but grows from the skin, just as a wart might grow, and when the skin is removed from a rhino's head, the horn comes off with it.

No one in this part of the world would dream of asking at a drugstore for a dose of powdered rhinoceros horn as medicine, but in many parts of China it would be quite an ordinary request. For centuries it has been considered an infallible cure for fevers and a source of strength in old age. It is so much in demand there even to-day that rhino horn has a commercial value in the East of about a hundred dollars an inch. Its medicinal powers were formerly believed in elsewhere, as well. In 1590, when Pope Gregory XIV was dangerously ill, the Prior and Brotherhood of the Monastery of Saint Mary of Guadalupe, of Spain, sent the entire horn of an Indian rhinoceros to His Holiness, with prayers for his recovery. The doctors cut off the tip, ground it to powder and administered it as a remedy to the dying patient. The remainder of this historic horn is preserved in one of the great museums of New York.

The faith in the healing qualities of rhino horn led

to another widespread belief that cost many a rhino his life. In bygone times, when rulers won their power through trickery and violence, their enemies often took their revenge by assassination. In India, and in Europe as well, there was always a secret conspiracy brewing at the courts of nobles and kings. Every tyrant was attended by a bodyguard. But though he was thus protected from open attack, he was never safe from the danger of poison. No matter how sumptuous the banquet, there was always the fear of what might be lurking in his dish or cup. Every prince had his corps of tasters, whose duty it was to sample the viands and wines set before the master.

There came a time, however, when a substitute was found for the wine-taster. A belief sprang up that a cup made of rhinoceros horn would instantly betray the presence of poison in the drink by showing fine drops of moisture on its outer surface. How and when this belief originated, no one knows, but it had a convincing hold. Noblemen and rich merchants discarded their handsome goblets of gold, and demanded drinking vessels of rhino horn. Hunters were despatched to pursue the beast through swamp and thorn for the sake of its horn, which was cut into cup shape and polished to a high luster, or sent to the subtlest craftsmen for elaborate carving.

The custom of using such drinking goblets spread through many lands and endured for many years. Wealthy travelers carried their own cups with them as a safeguard against the stealthy mixing of poison into their wine at some lonely inn where they might stop. A clever highwayman might disguise himself as a gentleman and make friends with a rich merchant over the supper table, and it would be an easy matter to drug his wine and rob him of his money-bags as he lay stupefied or dying, but as long as the traveler had his trusted cup to warn him, he felt himself protected.

The reason for this confidence in rhino horn is unknown. Recently, a group of chemists performed an experiment to test its alleged poison-detecting powers. They collected a number of such cups, set them in rows on their laboratory table, and filled them one after another with a variety of poisons, deadly and mild. Nothing happened. Not one of the cups "sweated," or showed in any way that there was anything inside more potent or more harmful than water.

Between the long-continued demand for his horn and the steady clearing of the land, the Indian rhinoceros declined heavily in numbers. Formerly, he ranged over the whole peninsula, flourishing wherever there was marshy lowland and leafy bush, but now he is to be found only in the Assam region, very sparsely in Nepal, and it is thought there are a few in Siam. The other two Asiatic rhinos are almost extinct.

The Sumatran no longer exists on the island that gave him his name, but must be sought in the untouched high forests in the interior of Borneo. He is the smallest of living rhinos, and the only one with a coat of hair. In his early years, his skin is covered with a fine, reddish down. As he grows older, this fuzz thins out, and eventually it disappears. His skin is minus the coarse tubercles that ornament the Indian rhino, and is much less rigidly marked in slab-like divisions. He is the only one of the three Asiatics with two horns. They grow one behind the other, several inches apart, and the front one is the longer of the two. In the baby, the rear horn is entirely unpointed, and looks like a flat metal button fixed into the top of his nose, but it shapes into a stumpy cone as he grows up.

The Indian giant rhino's single horn is round at the base and shaped like a huge thorn. It is usually between six and twelve inches long, though in rare cases it may measure as much as twenty. He puts it to practical use as a fender in forcing his way through thick growth, and, under rough and constant rubbing and scraping, the horn becomes scarred and nicked, especially at its base. Sometimes, in a headlong rush against a tree stump or rock, the horn may be ripped off close to the skin, but the damage is not permanent. Such a wound would cause bleeding and severe pain, but, like any other skin wound, it would heal and the horn would grow back again.

The Javan rhinoceros is also single-horned. As he is much smaller than the Indian, his horn is proportionately shorter, and sometimes it is no more than a bluntly pointed knob. Though all other rhinos show horns in both sexes, the Javan cow rarely has one.

In the last year or two this rhino has received more notice than ever before, because it has become known as the rarest of all the earth's large mammals. A famous English sportsman, who was interested in collecting the rhino family complete in all its five forms, went off to the Orient to try for a Javan, the only one lacking in his collection. He spent months scouring the thickets of the wild jungle where this rhino was supposed to exist, but found never so much as a hoofprint. Feeling that he could not devote the rest of his life to looking for this needle in a haystack, he gave up his personal search and sailed for home. Before his departure, he presented his case to the Game Warden of the District of Perak, in the Federated Malay States, where the rarest of rhinos was said to survive. As his interest was scientific, it was arranged that native jungle scouts should continue the hunt.

A year went by, and finally word came that tracks of a rhino had been found. Some of the beaters said they had even caught a glimpse of the beast. It seemed there was a last, single individual left, and the question was raised whether the one remaining representative of his kind should be killed. At last it was decided that if the creature died in the wild, it would be the end of his race anyway, and that, although it was a pity to shorten his existence, it was worth while for the sake of science. Meanwhile, the effort to trace the last of the Javans went on, and it finally led to the gratifying discovery that he was, after all, not the very last, but one of two or three.

This rhino is a little less hideous than his Indian and Sumatran kinsmen. The plates of skin over his shoulders are oval in shape, and the tubercles are very tiny, but the skin is creased and wrinkled in a network of fine, cracklike lines, in the confused pattern of a crazy mosaic.

The long search for the Javan rhino has benefited all others, for it has helped to stimulate interest in a more rigorous protection of them. A new society in the United Provinces of India is hard at work in their behalf, trying to persuade the rajahs and other influential natives to put a ban on free and easy hunting, and to have laws passed limiting permits to shoot rhinos, especially during the night hours, when the animals are out in the open, on their way to salt licks and drinking pools.

In Africa, too, rhino life has been made safer by stricter laws. There the race of rhinos was in no serious danger from the race of man until recently. As long as Africa was unacquainted with the gun, her wild creatures throve in huge numbers. The blacks had only their spears and arrows, and though they used these weapons freely, the animals held their own. The white man brought his far-seeing telescope and his long-range gun, and their introduction was the beginning of the end.

At one time rhinos were to be seen anywhere in plains country and thin forest in Africa, from the Cape of Good Hope to the lower edge of the Sahara. Now their free range is limited to Tanganyika, Kenya and northern Uganda. During the early years of the white man's invasion of rhino country, enormous slaughter took place. They shot the rhino for his meat. They shot him for his hide. They shot him for his horn. And some of them shot him just for the fun of it. Rhino meat is as good as beef, to some tastes, and although the huge brute is largely bone and muscle, a single rhino feeds a great number of men. The hide makes strong, stinging whips. When dried, it becomes so hard that it can be employed as resilient shields for the spear warfare of savage tribes. The white man's hand brought the touch of commerce, and new uses for the hide and horn were invented. The hide, when stiff, can be polished until it gleams like tortoise shell. Ornaments, table tops and all sorts of curios were fashioned of it. The horn was sawed into knife handles and knickknacks, or shipped off to the Orient to supply the demand for that sort of medicine.

At first the hunters' chief interest was in the horn, for the longer it was, the more money it would bring, or, if not taken for sale, the more imposing trophy it made. They hardly noticed anything else about the animal, and if they took the pains to study his habits, it was only with a view to more successful hunting. They knew, in a vague sort of way, that there were two kinds of rhino in Africa, and spoke of them as "the white" and "the black," though both were about the same dull gray in color. To the natives who worked for them in camp and on *safari,* the most important difference was that the white rhino is much larger than the black, and that his meat is fatter and better flavored. After a successful hunt, the porters always enjoyed a feast of roast rhino, and celebrated the occasion with dancing and singing. But their songs were not always cheerful. Sometimes they chanted lustily, praising the meat and thanking the rhino for the excellence of his flesh. At other times, they sang a song of reproach. "You are a stingy beast," they chorused. "You do not feed us well." It was always the black rhino that provoked their discontent.

Both African rhinos have better-fitting hide than any of the Asiatics. It is as thick and almost hairless, but it follows the lines of the body, more or less. The African looks less like a badly pieced puzzle than they do. There is a heavy fold at the thigh, and one at each shoulder, but it has nothing of the shield-like rigidity that suggests armor plate. Besides, it is quite free of the lumpy tubercles that disfigure the Indian, and of the criss-cross lines of the Javan. And there are no rolls encircling the neck.

His horns are long, and have the shape of gigantic, black carrots. The front horn is usually much the longer, although once in a while a rhino is found with horns of equal length. They weigh in the neighborhood of forty pounds. The front horn may be anywhere from a foot to a yard in length, or even more. A rhinoceros cow was once shot with a front horn measuring five feet and two and a half inches, which is about the height of a young person. From time to time, men have brought down rhinos with three and four horns, and one hunter reported one with five. Such curiosities occur very rarely.

As a general rule, the horns of the cow are shorter and more slender than those of the bull rhino.

Africa's two rhinos, or *kifaru,* as the natives of the east coast call them, differ very much in appearance and habits. The common, or black, is a bit less bulky, and

HEAD OF AFRICAN, OR "HOOK-LIPPED," RHINO

the shape of his head is like that of a tapering cask. His mouth is narrow, with a flexible upper lip that runs into a point. It is shaped like a triangular flap. This rhino feeds on twigs and shrubs, which the gripping muscles of the upper lip clutch firmly. From this feature, he has been named the "hook-lip."

The white rhino's head has much more rectangular lines, and resembles a box more than anything else. His lips run in a straight, horizontal line, about twelve inches across, and the lower one is reënforced with a horny band on the outer edge. He feeds on grass and low-growing weeds, and the hard rim of his lower lip enables him to clip his pasture close, with the effect of a lawn-mower. He is fastidious in his tastes, and leaves bitter weeds untouched. His broad, straight mouth has given him the name "square-lip." He is the only one of the five rhinos with this lip formation. His neck muscles rise in a bulky ridge, and this, too, is a special feature of his own, and makes him easily distinguishable from all other rhinos. The development of these tremendous muscles is probably connected with his manner of feeding, for whereas the hook-lip and the Asiatic rhinos are all browsers, the square-lip is a grazer and must lower his huge head straight to the ground when he eats, and this means powerful and elastic muscle manipulation.

Although the names "hook-lip" and "square-lip" are so vividly descriptive of the two African rhinos, they are not as much used as the familiar "black" and "white." To the first careless observers, it seemed that there actually was a difference in color marked enough to distinguish one from the other. But it is really only a matter of mud! It happens that all rhinoceroses are extremely fond of bathing, and the dirtier the bath, the better they like it. In spite of the thickness of their hide, they feel the heat of the sun and the sting of the millions of insects that cling to them, gouging out a living from their bodies. A mud bath not only cools the skin, but coats it with a layer that keeps off both insects and sun, for a time, at least.

The square-lip seems to need his mud-soak more desperately than the hook-lip, and wherever he is found, there is sure to be a mud-hole somewhere near by.

Though it means a trudge through sharp, jagged plains-grass and thorn bush of several miles, he must have his daily dip. A rhino wallow is thick with cooling water weeds, wild lilies and swamp reeds, and is often shaded with marsh palms. It is a paradise of ooze, where a tortured *kifaru* can sink in blissful comfort, his beady eyes closed, and forget the woe of a smarting hide. When he drags his dripping body out of the slime, he carries a load of wet clay off with him. If the earth happens to be whitish in color, the rhino emerges with a coat of whitewash. He is likely to top off his bath with a roll in the dust, and this adds to his creamy appearance. When the sun has dried the caked mud and powder, he begins to itch all over again. He may find relief by rubbing his great sides against a convenient anthill. The African plains are dotted with huge anthills, which rise like mounds among the grass. Again he takes on a dusty pallor, but when the dust has blown away, he is his former gray self once more. Sometimes a "red" rhino is seen, in a region where the earth happens to be so tinted. Or, if he has been in a wallow that is stale and overgrown with slime, he may emerge a handsome shade of green, so that no color distinction really applies.

The square-lip was formerly known only in South Africa. Under the rifles of the hunters and settlers he almost disappeared. When the government authorities realized how near extinction he was, they decided to save what few survivors there might be. They set aside the tiny district of Umfalosi, in Zululand, as a protected reserve, and no hunter was permitted to shoot a white rhino within its boundaries. At the time the rhino community there numbered a dozen, but now the population has more than doubled.

In 1900, to the surprise of every one interested in

rhinos, a hunter named Major Gibbons announced that he had seen square-lips on the west bank of the Nile, fully twelve hundred miles north of Zululand. Later, other men, among them Theodore Roosevelt, reported frequent meetings with them in this northern stretch. There are between two and three thousand of them in

HEAD OF AFRICAN, OR "SQUARE-LIPPED," RHINO

the savannah land (you may be sure there is wet, wallow ground wherever the white *kifaru* roams) between the northeastern edge of the Congo Rain Forest and the papyrus beds of the West Nile. None exists on the east bank.

In spite of his size and strength and his powerful, pointed horns, the rhino is not a wanton killer. His chief occupations are feeding, sleeping and wallowing. He has no animal enemies. He is too big for the lion to tackle, although once in a while an unwise and unfortunate calf strays away from its mother and falls victim to one of the Big Cats. Elephants seem to show a dislike for the rhinoceros, especially in India, where they have been used as mounts in rhino hunts, and a few have been found with marks of wounds that seemed to have been the work of the slashing horn. But there is only this evidence to go on. No one is known to have seen a combat between these pachyderms. Rhino bulls occasionally engage in duels, and one man who happened to witness such an engagement found it anything but a pretty spectacle. The two animals tore after each other in a circle, snorting and puffing, aiming their horns, like swordsmen, and inflicting terrible wounds. The Indian rhino relies on his giant incisor teeth for his fighting, but the African is not equipped with tushes and his method of attack is the horn thrust.

Fights are not frequent events in rhino existence, yet the *kifaru* has an ugly reputation. The black, especially, is said to hold the animal record for bad behavior. He is absolutely not to be counted on. He may entirely ignore your presence, and go on dozing or feeding. He may trot off quietly in another direction. Or, he may, the instant he catches your scent, come at you in a savage, head-on rush. Once started, he crashes straight on, deaf to all noise, blind to all obstacles. From the mass of his body and his ponderous short-legged build, no one would suspect the rhino's racing powers. He surges forward like a solidified gray cyclone. Everything goes down like straw before the violence of his strength and the impetus of his weight. The first symptom of his alarm is the forward pitch of his ears, but the real sign of a coming rush is in his tail. When a rhino means murder, his tail screws up into a tight little curl. A charging rhino is like a boiler-engine run amok, shrieking and snorting and plunging into and through everything on the path of his rampage.

His rush would be fatal, if it were not for his one great handicap. His sight is bad. His nose and ears tell him there is something strange and alarming about, but he cannot clearly locate it. He is so near-sighted that he is literally blind to what is going on around him. Straight ahead, he can see, but not to either side. His eyes are set up near the arch of his nose, about halfway the length of his head. They are ringed with thick folds of skin, which furnish protection against the thorn and scrub through which he travels, but they apparently obstruct his vision. The horn, too, hampers his sight. His squinting eyes are small, not over three-quarters of an inch across. His nostrils are six times as large.

If a man is unlucky enough to be trapped in the path of a raging rhino, he may be impaled on the terrible front horn, or trampled to death under the pounding feet. But if he is quick enough and skillful enough and has his wits about him, he can escape the crashing whirlwind simply by stepping out of its way. The rhino lunges straight on, never noticing that his target has disappeared. At the very climax of his rush, he may stop abruptly, as though his fright or temper had suddenly cooled, and trot off in another direction. On the retreat, he may change his mind and turn back for another furious rush.

Sometimes a perfectly placid rhino is overtaken by a frenzy when there is no apparent cause. It may be nervousness or ill-temper. Perhaps he is bothered by pain. After all, wild animals have toothaches, and stomach aches, and most of the upsets that we have. Perhaps he merely gets the fidgets. Whatever it is, he may start up all of a sudden out of a snooze and go catapulting across the plains, crashing into anthills, barging through bush and tree clumps, unchecked until he wears off his mood of violence.

BLACK RHINO IN AN UGLY MOOD

The white rhino is less given to such spasms, but the Africans say there is a devil at home in every *kifaru*.

Except for these occasional moments of insanity, the

rhino is a model of good conduct. He lives at peace with his fellows, generally going about with his own family and minding his own business. The parents are often accompanied by a calf on their way to the wallow. The baby walks ahead of his mother, and she keeps him to the path by prodding him with the end of her snout, so that he early learns it is rhino custom to follow the beaten track.

During the day the *kifaru* takes things easy. Having spent most of the night awake, he is tired and takes his sleep in brief naps, waking up now and then to eat. He lies down when and wherever he feels like it, in the shade of a tree or anthill if there is one at hand, otherwise on the open, broiling plain. He sleeps in safety, for every rhino has his bodyguard in the little tick birds that perch on his head and back, feeding on the insects that infest his hide. He is their banquet table, and they are welcome guests. Whether intentionally or not, the birds repay his hospitality fully. While he sleeps, they are wide-awake, busy exploring his person. If anything frightens them, they set up a chorus of chirps and terrified cries, fluttering up nervously and flying over the rhino's head, beating their wings and screeching until the sleeping giant awakes. They are probably scared on their own account, but their noisy panic acts as an alarm signal which the rhino unfailingly heeds.

A rhino can jerk himself to his feet with surprising speed, and, when he is disturbed, he usually stands up to investigate. Once in a while, he sits up on his haunches for his survey. Theodore Roosevelt tells of his amazement at seeing one of the huge brutes sitting up dog-fashion, which is very unusual for so big an animal.

The *kifaru* is not an adventurer. When on the way

to drink or to the wallow, he rarely strikes out on a new route. He treads in the broad, deep trails cut by his fellows, sniffing as he goes to pick up what news there may be of others of his kind. These trails, traveled by thousands of rhinos for thousands of years, scar the land. Like the bison of America, the African rhinoceros has paved the way for road-builders and engineers who plot the lines of travel by motor and rail as, more and more, the continent opens to civilization. More than one rhino has been startled out of a sound daytime map by the chugging of a car as it cut across the plains. Rhinos and autos can hardly occupy the same territory, and before very long the rhino will have to give way.

13

THE WOLF

In the chaos of the retreat from Moscow, Napoleon's army—that famous Grande Armée, once the pride of the Little Corporal's heart—presented a picture far remote from anything we think of to-day in connection with the movement of troops.

Over their weather-beaten uniforms, so unsuited to the bitter Russian winter, the soldiers had wrapped any woolen garment they happened to find in their looting of the Russian countryside. From the stables of rich Russian nobles and merchants they took elaborately fitted, silk-upholstered coaches, to which they harnessed their half-starved army horses, and into which they stuffed all the loot they could collect. Their progress was the extreme of disorder, encumbered by large objects they had taken from the houses of Moscow, now in flames behind them—enormous statues, ornaments, bags of heavy golden coins. Every bridge was a scene of confusion. Officers vainly tried to enforce some order of march; bridges broke under the helter-skelter crowding of men, carriages and the unwieldy guns they had dragged with them all the way from France to Russia.

On the sides and in the rear, as they retreated, the Cossack on his sturdy, swift pony constantly harassed the weary foot soldiers by surprise attacks, riding off as swiftly as he had appeared.

Little by little the horses died under strain and starvation, for in spite of the golden loot, there was no food. Napoleon himself left his ragged army to hurry back to France. He had gone to raise another army for the next winter. All discipline went with him. The men were panic-stricken and footsore. They grew gaunt and gray. Their heavy golden treasure lay by the wayside, discarded and covered with the heavy snow. By the time it reached the Polish frontier, the army was a collection of scarecrows in headlong flight.

And behind the army came a menace even more terrifying than the Russians—a lean, yellowish-gray creature, as hungry as the men, and even more desperate—the Russian wolf. It sped silently behind the forlorn men to the Polish border, sweeping down into Germany where it took up a fatter living in the fine German forests near settlements whose sheep and cattle fell prey to its savage jaws.

Thus Germany, which with much effort had been freed from this marauder of the night, was once again stocked with the wolf, and a particularly strong, savage brand of wolf.

This animal, from time immemorial, wherever it has roamed has been a scourge to the inhabitants. Just as the Puritans protected themselves against the savage Indian tribes by banding together against their surprise attacks, so the inhabitants of wolf-infested lands tried, and still try, to protect themselves from the packs of this savage animal who is a menace not only to their flocks but to them and their children.

Through the vast tracts of the Russian and German forest country, packs of the creatures will follow sleighs for miles. In the dark night the sound of their hungry, high-pitched, insane howling carries into the lonely little Russian villages and the outskirts of towns. And in the morning despoiled sheep pens tell the story of this thief in the night.

Some of the organizations against the wolf still exist. In Switzerland, the men voluntarily banded into Wolf Clubs, and elsewhere the menace was so great that the government took a hand, as in Brittany where the Grand Louvetier is still a government official. For even where the wolves have been exterminated, a hard, foodless winter will bring packs or single animals across from the wilder parts of mountainous Europe to easier sources of food.

It takes a thief to catch a thief, for their own near relatives, the dogs, are used in pursuit of the wolf. Compare the dog that is used as a protection to man—the police dog, so trustworthy and faithful that he can guide the blind, or the Eskimo dog so useful in the Far North —with this other canine who is such a danger to man. And whose nearest relatives among the dogs are just those two!

The outward resemblance to his near dog relatives is very strong and if you met a wolf being led along a street, you would probably think he was a police dog. Like the police dog the wolves are long-bodied and long-nosed, a strong, lean, high-haunched animal. The long hair is longest in the rather larger wolf of the North. But the chief similarity between wolf and dog is the fierce canine teeth. The color varies. The commonest color is yellowish-gray, or brownish-gray, but there are white wolves and black wolves. In good condition, the wolf is a beautiful animal except for its expression which is a mixture of nervous fear and extreme ferocity. But it is more often in the unbeautiful state of having a "lean and hungry look."

Extreme speed enables the wolf to hunt as he does,

not lying in wait for his prey and creeping up upon it, but hunting in the open, counting on his ability to overtake his prey. Men who go out to hunt him have to use his own tactics—speed. So difficult is it to outwit or overtake him that wolf-hunting has become not only

WOLF

a defense measure but one of the most exciting sports. In the days of Ivanhoe, England was haunted by wolves and not only was wolf-hunting a favorite sport, but a very necessary measure. So usual was the menace that refuges were built in which travelers might seek shelter if wolves attacked them. Several English kings tried their best by royal edicts, offers of reward, and other measures, to get rid of these pests, and finally succeeded in driving the surviving ones into the more desolate parts of the country. When the last wolf was destroyed in England is uncertain; probably not until the late 1700's.

In America, however, it is another story, for here the wolf has never become extinct. It still roams the Rocky Mountains, the Pacific coast and northern Canada. Black wolves still exist in the Florida Everglades, and in the Arctic there is a very beautiful white wolf, pure white except for a black tail-tip.

They are difficult to exterminate, both because of the wild and inaccessible haunts they frequent and because of their fecundity. In former times they used to occupy a much larger area in America, but increased settlement and destruction of their food—other game—has cleared them out completely from large areas where they used to be. However, only recently they have suddenly reappeared in an area where they have not been for a hundred years. Just within a few months, wolves have begun to prowl through the Adirondacks; farmers report hearing the howling at night; hunters report seeing packs of the gray marauders, and one official in an Adirondack town reported that wolves in his town were so numerous that they were rapidly depleting deer herds and killing game birds and rabbits.

Where there is loneliness, there the wolf has stayed in the isolated parts of both New and Old Worlds. In the Pyrenees there is a black wolf; there is a small wolf in the Ardennes and a rather larger one in Germany. In Turkey and Italy there is a tawny wolf, while the lonely stretches of Poland and Russia, Siberia and Central Asia are ideal refuges for packs of this swift animal with its strong jaws and cruel teeth.

The Indian, Central Asian and Russian wolves are very fearless and fierce toward man, but the American wolf, although ruthless in his attack on other animals, shows more inclination to fight shy of man. Even when they are cubs, all jaws and paws—for, like tiger cubs, they "grow to the head"—their playing is rough and snarling and as they get older, to this fierceness they add a keen intelligence—a shrewdness familiar to us in all tales of wolves from Æsop's Fables to the story of Little Red Riding Hood.

This intelligence leads to a mutual helpfulness in pursuit and capture of their prey. Except in the mating season and when the young are being cared for, they hunt in a formed pack. When the young are being cared for, the parents abandon the pack; the mother stays at home in a den among rocks or in a burrow in the hillside to care for her cubs, while the male wolf hunts and brings in food for the family.

But whether hunting alone for his mate and young, or in packs, the wolf's ravages among cattle and sheep are cruel and unceasing.

On the wide, brushy plains of the United States and the table lands of Mexico, there is another wolf. This is a smaller, slinking animal, and one whose disposition is just plain mean. In the West, you see these yellowish-brown creatures trotting through the sagebrush, and the cowboys will describe a particularly unpleasant person by saying he is "as mean as a coyote." The more towns there are in their district, the more sheep and poultry fall to their snapping jaws. A price has been put on the coyote's head, but he is so swift and so cunning that he is a pretty good match for trap, poison or gun. The coyote does more solo hunting than the other wolves. He is always sneaking in around camps trying to steal chickens, bacon or other supplies, but failing to get such food, he will eat cactus or even the sweetish mesquite bean.

In contrast to the wolf of the hot, arid deserts, the white Arctic wolf—*Canis tundrarum* (dog of the tundra) —wanders around the borders of the Arctic coasts of Canada and Alaska and the islands as far north as the coast of Greenland. Up there, during the short summer, swarms of wild fowl furnish an addition to the regular food supply. That is when they feed most heavily and store up a reserve supply of fat for the sunless season. When winter comes, they chase lemmings, Arctic hares, and sometimes even the white fox. In the cold dusk of the winter months north of the Arctic Circle, the wolf pursues any food he can get. He does not stop at smaller prey; he hunts the caribou and the musk ox and even cattle—all these of course very much larger than himself. But these animals, although larger, are poor matches for this swift white creature of the snows who follows them relentlessly and tirelessly, and as relentlessly kills.

The Eskimo dog, who with the reindeer shares the task of being the horse of the Far North, is closely related to the wolf. These dogs are sometimes crossbred with wolves to give them increased strength and speed. Such hybrids of wolf and dog do not bark—they howl!

Across the straits, in Siberia, where food is even scarcer, the wolf is even more terrible. Near settlements they flee along the roads in terrible packs, pursuing sleighs. Yearly from remote villages in Russia and Hungary patients who have been bitten by these hungry creatures are sent to the Pasteur Institutes, for rabies—the disease that makes a dog "mad"—is very common among wolves, and seems to be getting even more common in recent years.

It is a far call indeed from the lonely stretches of

Siberia to New York's elegant Park Avenue. But on Park Avenue nearly any day one can see beautifully dressed women strolling along accompanied by one of the haughtiest, most aristocratic-looking dogs in the world, a thin, delicate, high creature of greyhound build with straight long hair—the Russian wolfhound. But delicate and haughty as they may look when they are on a leash and forming part of a well-dressed lady's costume, these same dogs have been so christened with good reason. The borzoi, or wolfhound, is such a famous wolf hunter that it has been imported into America and is used to course wolves in the western states, for sport such as Theodore Roosevelt enjoyed. These dogs are reckless, speedy hunters, as speedy as the fastest race horse, and although the ones over here are not quite as savage as those in use in Russia, they are expert at overtaking and holding the wolf. But even the fleetest find it difficult to overtake a full-grown wolf.

On the steppes of Central Asia, the Tartars have devised an aid to the dog in wolf-hunting. There the wolves do not hunt so much in packs; they are more apt to go singly, but the damage they do is immense. The nomad tribes of this region are chiefly occupied in breeding and raising horses and cattle. They have no fixed home, but follow the grazing grounds. They spend a large part of their life on horseback which puts them in position for immediate pursuit when they spot a wolf near their herds. But the dog is not enough as an aid. There is another savage hunter that they can call in to their help—a bird, the huge, fierce eagle. These birds the Tartars and other wild tribes capture and use as very civilized hunters in France and England use another bird, the falcon. But the eagle cannot be carried like the falcon, on the hunter's arm. It is too heavy. So these horsemen make a wooden rest beside their saddle and on this they rest their arm. When the wolf is sighted, they loose the eagle from its strap, and the big bird flies straight at the animal and attacks it savagely. With the big wings flapping about his head, the cruel talons and the sharp beak all in use against him, the wolf becomes panicky and in his confusion loses sight of the approaching men who can then get close enough to kill him.

In Siberia, the wolf's fondness for pork has been given full consideration in a very novel and hardly sporting way of hunting. The driver of a Russian sleigh usually sits on a rather high seat in front of the passengers. He takes his place, and behind him sit the men, armed with guns. But they have with them another weapon—a small pig! This small animated implement they induce by proddings and pinchings to squeal as loudly and as continuously as possible, hoping that the wolf in the nearby forest will hear it and be lured to approach. Behind the sleigh the hunters have attached a large bag of hay which drags along the road. When the wolf, enticed out of his hiding place by the pig's complaints, approaches the direction of the squeals, he sees this fattish object trailing along behind the sleigh. He leaps upon it to seize what he imagines is the pig—and is promptly shot by the passengers of the sleigh.

Universally the wolf is a hunted animal, the symbol of ferocity to which hunger may drive any starving animal. However, one writer has given quite a noble picture of this beast. Rudyard Kipling in his Jungle Book gives the wolf quite a measure of grandeur. Æsop granted him shrewdness and the old legend of Romulus and Remus, who were nourished by a she wolf, shows the wolf in an aspect of obligingness at least. But far more numerous than Kipling's wolf, is the wolf which appears in stories—even in whole books—in what is unfortunately the truer and more characteristic aspect—a strong, lean creature of the wild, treacherous and thieving and a menace to better, finer and more useful animals.

14

THE DUCK-BILLED PLATYPUS

In the laboratory of a museum sat a zoölogist, staring at the skin of a small animal; a nice furry skin, soft and dark brown, not particularly unusual. But what his eyes were fixed upon was not the fur. It was the animal's mouth. He felt around its edges where it joined the body, suspicious for a moment that some one was playing a joke on him, or that another "mermaid" was in his hands.

The animal was reported to have come from Australia, but it had come in on a boat on the Eastern run, and sailors on the Oriental runs often bring back with them "mermaids." They are brought into natural history museums almost every year, sometimes for explanation and sometimes just as curiosities. They are very different from the mermaids of poems and legends, those beautiful, mythical water creatures with the faces of lovely maidens, long flowing hair, and glistening fish tails. These mermaids are far from beautiful, maidenly or glistening. They are usually only about two feet long and a very dark brown color, very wizzled and dried-up as to face and claw-like hands. Their tails, although fish tails, are hard and shriveled. In fact, close examination reveals that the clever fingers of some Chinaman have stuffed with straw the upper part of a small monkey and the lower part of a largish fish, and have sewn the two together with almost invisible stitches. And there you have a mermaid! Not at all alluring, but peculiar enough to be puzzling.

So the men who first saw the skin of this small Australian animal suspected at first that a new variety of mermaid had arrived on the scene. For attached to the front end of the small furry animal body was a large, flat duck bill!

But there were no stitches. It was quite obvious to the zoölogist that the bill really belonged to the animal. It was also clear to him that the creature was a mammal, but he had never before seen mammal or bird or anything else that was furred, had a duck bill and webbed feet. Still further examination revealed this creature as peculiar beyond his wildest nightmares, but for the moment his chief concern was to name it and make its existence known to other scientists.

When a scientist finds an animal that has never been accurately described in any book, as far as he can discover, he sets out to give it a scientific name, and to write a description of it, usually also drawing its picture. This description he publishes so that people all over the world will know that the animal exists, where it was found and what it looks like, and will be able to refer to it by a name which is universal.

Usually such an animal already has a native name, or even several different names. In this particular case the animal was already known in Australia as the "duck bill," "water mole," and by several native names.

Often an animal has very different names in different parts of the same country and in different countries. This causes great confusion in any article written about the animal, especially if no pictures accompany the article.

If an English-speaking person is talking about the animal we call the horse, he calls it "the horse," but a Frenchman calls the same animal "le cheval"; a German calls it "das Pferd"; an Italian, "il cavallo," and a Russian calls it "loschad." You might open a French book and see a drawing of the inside of a hoof, with the label beneath it "le cheval." But unless you read French or knew the subject very well, you would not know that this was the hoof of the animal you call the horse. However, if under the picture there had also appeared the words *Equus caballus,* then, without a doubt, you would know that here was the common horse, even although the vernacular name above the Latin words were to be in Hindustani. It is for this reason that scientists give animals what we call a "scientific name," that is, a name which belongs only to that one animal and is the same all over the world. Thus the horse, scientifically in every country in the world is *Equus caballus,* and *Equus caballus* cannot mean anything but the horse.

Customarily these scientific names are made from Latin or Greek words, and some years ago even the descriptions of the animals were written in Latin, or the description in the author's native tongue was preceded by a condensed description in Latin. But now that fewer people read or write Latin, and many cannot read it at all, that custom has disappeared. But the universal scientific name remains.

Such a name is Hippopotamus, composed from the two Greek words *hippos* and *potamos,* meaning "horse of the river." Because there is usually more than one kind of the same animal, the scientist gives each kind a different second name. The first name, for instance, *Equus,* is called the generic name or name of the genus or general group to which the animal belongs—the horse. The second name is the specific name or the name of the species or special kind of animal in that group. *Equus przewalskii* is the wild horse of the Asiatic desert; *Equus onager* is another member of the horse tribe—the wild ass of Asia.

All animals have two names, for even if only one of the kind is known at the moment, there is always the possibility that another variety may be discovered. New varieties and even completely new animals are constantly being found in little-explored places—in inaccessible mountain ranges, tropical jungles, the deep sea or the vast deserts.

It is very important that no two animals be given the same name. To prevent this, every year a book called the Zoölogical Record is published, listing all the new names given during the past year. But of course sometimes people cannot get at this book, or one animal is named twice in the same year by different people. In such a case, the name first given is the name that stands. Unless, as sometimes also happens, the name has been previously given to some other animal. In this case a new name must be given as soon as the duplication is discovered.

Both these things happened in the case of the platypus.

Dr. George Shaw, the zoölogist who held the small skin in his hands and wondered if another mermaid had been concocted for his benefit, gave it the first name, *Platypus anatinus* (*platus*—flat; *puos*—foot). Within a few months another skin was sent from Australia to a German museum, and the German anatomist who examined it, not knowing that Dr. Shaw had already named one, called it *Ornithorhynchus paradoxus* (*ornithos*—bird; *rhynchus*—beak).

When the two descriptions were published, it was first found that before being called *Ornithorhynchus,* the animal had been named *Platypus*. Then it was found that the name *Platypus* had already been used some years previously to designate a certain beetle! So the duck-billed animal is correctly known, scientifically, as *Ornithorhynchus*. However, such excitement had been caused by the appearance of this skin in the British Museum, that people all over the English-speaking world were calling the animal platypus as if it were a common name, just like "duck bill." So although scientifically it is incorrect, the original scientific name, *Platypus,* has become the every-day name of *Ornithorhynchus*.

You need not bother to remember the second name, for so far there is only one *Ornithorhynchus* in the world, and in only one place in the world—Australia.

This animal is of such ancient lineage, that, if it had followed the usual course of zoölogical specimens, it would have been a fossil long ago! As it is, one might describe it as a living fossil—the living descendant of a most ancient race—the Monotremes, the lowest order of living mammals.

This animal, about the size of a small cat, is characterized by the flatness from which it derives its name. Its bill is flat; its large, webbed feet are flat; its bulky tail, furless below, is flat. Its whole small, wide body is flat. When held in a hand, the animal is limp and helpless-looking, both because of this flatness and because its skin is so loose. A deceptive helplessness—for it has all the abilities of a first-class contortionist.

Mr. Harry Burrell, who has had more opportunity than any other naturalist to observe the platypus in its native home, says that "when lying fully extended on its back, the platypus can, by placing the lower portion of its bill on its breast, and without raising its head to any appreciable extent, double itself ventrally until its head passes its tail, and that pliable member is itself doubled until the creature becomes normally righted on all fours, dragging its tail behind it. This act it can accomplish in a tunnel equal in circumference to the performer so doubled!"

THE DUCK-BILLED PLATYPUS

It can fold its tail over its head, or in any other direction to the bill. It can turn its head at a right angle to its body; it can walk in either direction without turning around, and, without rising from the ground, can lengthen itself as much as six inches!

At the same time that the skins of this creature were journeying to various museums, a persistent rumor came from Australia that the animal not only had a bird's bill, but that it laid eggs like a bird. This news was of utmost interest to zoölogists, for mammals do not lay eggs and the animal was undoubtedly a mammal.

The platypus lives in water and in the banks of rivers and is exceedingly timid and secretive. Between its furtive habits and the complete inaccuracy of reports obtained from native tribesmen who saw the platypus frequently but whose stories did not match, it seemed impossible to find out whether this rumor about egg-laying was true or false.

Natives were hopefully questioned. And two natives from the same small village, visiting the same stream of water, would insist equally firmly, one that the platypus laid eggs; the other that "no egg; pickaninny make tumble down" (the young are born alive).

One enterprising investigator drew a picture of a nice large egg, a round egg, and showed it to a native. The native eagerly claimed to recognize the platypus egg. The investigator then drew a picture of an oval egg and showed it to another native. And the other native eagerly claimed to recognize it as a platypus egg! So the investigator threw away both pictures and also the idea that he could get any accurate information from native observers.

Then a series of "platypus eggs" began to appear, particularly as correspondence in newspapers and magazines made public the search for the egg of the elusive creature.

One naturalist received a letter from a settler in Australia announcing that he owned a platypus which had been captured in one of the small rivers and presented to him. This gentleman had put the animal in a cage, and wrote to announce that the morning after there were two eggs in the cage with the animal—white, soft eggs about the size of a crow's egg. At last, wrote the proud owner, a platypus had actually laid an egg.

Unfortunately no egg accompanied the letter, but from further correspondence and a description of the two eggs, it became all too clear that one of the proud gentleman's friends had made a night journey to the cage of the platypus and had placed within it two bird's eggs!

Then another report came to light amid much excitement. Again the platypus egg had been discovered. Its discoverer wrote that unfortunately, through ignorance of its scientific value, the egg had been destroyed in handling, but he sent in a very detailed description and an accurate drawing. So accurate was the drawing, that the disappointed scientist who received it had not the slightest difficulty in recognizing in it a very good picture of the egg of the common Australian long-necked tortoise!

Finally, an English zoölogist set out for Australia, determined to settle this question. A few months after his arrival there, a cable announced that the problem which scientists had failed to solve for over eighty years was at last settled. He had gotten a platypus egg and had proved absolutely that the platypus was its author!

But there was still a lot to learn about the secretive platypus. No platypus had been carried alive out of Australia and it was surprising for people who had never seen the animal alive to learn that the duck bill which hardens when the skin dries, is in life moist, softish, and flexible. This bill looks as if some one had pulled on a dark leather glove up as far as the animal's eyes where the wrist gauntlet of the glove lay loosely above and below, the top flap flaring up nearly to the small, bright eyes.

The whole muzzle is of naked skin and is remarkably sensitive, for it contains a mass of nerve fibers. This is the animal's chief organ of touch. Unlike many other mammals, it has no vibrissæ, or "whiskers," with which to feel, but instead, unlike any other mammal, it has this bill, covered with special tactile organs. The most sensitive portion is the front edge of the upper lip, in

THE BILL OF THE DUCK-BILLED PLATYPUS

spite of the fact that in burrowing this is the part first used to throw up the earth.

Through this bill occasionally comes the voice of the platypus. Its only relative, the echidna, appears to be almost mute. At best it emits a sort of sniffle. But the platypus has a low voice like a trembling snore, and young ones if disturbed will indulge in low growling.

The platypus relative, the echidna, does not look anything like the platypus and probably the two never meet, but anatomists find, by examining their body structure, that these two are related. Both lay eggs and both burrow, although the echidna burrows in dry ground, rapidly digging down beneath himself. This animal has a long, sticky, extensile tongue and is covered with spines, from which two characters he gets his native name, the Spiny Ant-eater, although he is not related to the true ant-eaters of tropical America. The echidna resents captivity and always makes strong and determined efforts to escape. Among the Australian native tribes, the "native porcupine" is a favorite food. They pack mud around the spines and roast him.

Inside the long platypus bill there are no teeth in the adult, although the young platypus has small ones. Instead of teeth, the animal has two horny rims on which it can chew. But it gathers in its food through compressing the bill inside into a sucking tube, through which it draws in the food. Some of the food is stored in pouches in the cheeks. With laborious effort the platypus sucks up from the river bottom and stores in these cheek pouches enormous quantities of—mud!

But if any one imagines they can feed a platypus on mud, they are very sadly mistaken. The problem of feeding this animal is the chief thing that has made its exportation practically impossible. This, and the difficulty of providing it with anything resembling its regular home.

More interesting to us than its smallish, dirty-white egg is the home where the egg is laid. Both male and female platypus spend part of their time in the water where they feed. But their rest and breeding periods are spent in and on the river banks, in their burrows.

There are two very distinct varieties of burrow, the resting burrow and the breeding, or nesting, burrow.

The resting burrow is a simple structure. It is a short tunnel into the soft earth of the river bank, often made under an overhanging rock or the large root of a tree. Usually one platypus, and only one, occupies a burrow, but sometimes two males peacefully share one. A male and a female never occupy the same burrow.

The platypus leads a very neat and tidy life. After his meal in the river he emerges from the water and proceeds to comb his wet hair, chiefly with his hind paws, but helping with his forepaws and his bill. Because his skin is so loose, he can pull around any portion of himself he wishes to get at, and work at it comfortably. For the season of his residence in any one spot, the platypus usually uses one definite place in which to make his toilet. This is generally a large, flat rock, near the water so that at the approach of anything alarming he can slide quietly off his toilet table into the concealing stream.

Sitting on his rock, he smooths his hair until it is comparatively dry and very sleek and shiny. He then retires to his resting burrow and takes a nap, so curled up that he seems nothing but a compact ball of brown fur. Very often you will notice a slightly onion-y smell about the resting burrow of a male platypus. This comes from a scent gland just in front of his shoulder.

The nesting burrow of the female platypus is an elaborate structure and she works very hard to build it. It is narrower in general than the resting burrow, but much longer and much more complicated.

As chief building tools she uses her bill, the strong webbed forefeet and the bulky flattened tail. The web of the forefeet projects far beyond the claws and in swimming and digging this web is spread out, although in walking the web is folded neatly back under the claws and the platypus walks on the claws alone. These are slightly convex beneath, thus preventing them from tearing the web.

The back feet are also strong and webbed, but not

as widely so as the fore. They have, however, a peculiarity of their own in the male animal. On the inside of the foot there is a strong, claw-like spur, and through its hollow interior the animal is capable of ejecting a poisonous fluid. This fluid, however, has not been proved to be fatally poisonous, nor does the platypus often shoot it out.

FEET OF PLATYPUS SHOWING WEBBED FOREFOOT AND SPAN OF HIND FOOT

With the end of its flexible, stout tail tucked tidily between its short hind legs, the platypus systematically begins her building operations by digging her hind claws firmly into the earth. Thus propped, she stiffens her upper lip, spreads her webbed forefeet and begins breaking away the dirt with lips and paws.

After she has made a hollow of a few inches, she rolls her body from side to side and slaps the floor of the tunnel with her tail to tamp the loose earth into firmness. Thus she progresses into the burrow she makes, feeling with the sensitive lip, breaking, shoveling, rolling and packing the earth; reaching out a front foot as far as the webbed claw will go and breaking new ground ahead of herself.

No earth is thrown out. The system seems to be to dig about twice as widely as the finished burrow will be when the loose earth is packed down. During this process there is sometimes a landslide and the front end of the platypus is snowed under. The platypus then finds it convenient to become acrobatic. With ease, if not with grace, she "advances backward," simply reversing the action of her hind legs, pushing with the front ones and alternating the movements of the hind ones just as she ordinarily does in walking on the forefeet. In the darkness and loose earth of a burrow, it is hard to tell which end of the platypus is which when it progresses toward you tail-on!

While making its burrow, as in swimming under water, the platypus closes tightly its eyes and ears. The eyes can be closed by the ordinary methods of shutting them; the ears, which are invisible to the casual view because there are no external ears, merely openings under the fur, may be closed by bringing together the edges. But both eyes and ears can be closed at once by bringing together the edges of the skin furrow in which they both lie.

The platypus breeding burrow, about a foot below the surface, arched on top and flat under foot, is usually long and winding, sometimes as long as fifty feet. It is always built well above water level and has only one entrance. The burrowing platypus shows a remarkable instinct in avoiding the burrows of other animals. She will get within a few inches of one of these, then turn in her path as if she sensed its presence. Again, she will dig down and carry on her own burrow beneath the other burrow.

At the end of her burrow, or within a few inches of the end, she carefully constructs her nest, sinking a round hollow of about a foot wide, a little below the level of the tunnel and lining it with a thick layer of leaves. How and when the platypus gets the leaves, grasses and twigs of her nest into the burrow is a matter of guess work. She probably takes them in at night.

Only one female occupies one burrow. The same burrows may be used in subsequent years, or they may be renovated and enlarged. One ancient and much-used burrow was discovered one hundred feet in length. But usually the platypus makes a new burrow each breeding season. As no one yet knows how many years the average platypus lives, we do not know how many times during her life she performs this task. She stays in the nesting burrow only while she is laying the eggs and until the young are past the nursing stage.

Before she lays her eggs she defends herself against possible invasions. In her tunnel, she makes blockades. At a short distance from the entrance, she begins to plug the tunnel at intervals with loose earth. To do this, she digs out a side tunnel from the main one, using this earth to plug off the main branch. These "pugs" occur more frequently as the tunnel approaches the nest. If the platypus is pursued into her breeding burrow, she will rapidly throw up behind her a series of pugs. It seems impossible that the female and young get enough air to live in the nests at the end of these long tunnels. Whatever air they do get must be the small amount that seeps through the loose earth of the pugs. She and the young apparently require very little.

Whenever she has to emerge from the nesting burrow, on her return she carefully reconstructs the blockades.

The eggs, of which there are usually two or three, are

dirty-white, unattractive objects, and the young platypus is also unattractive. Like the young kangaroo, it is very tiny and very bare, and it develops no fur for several months. This small animal obtains food from its mother in an odd way. Unlike other mammals who nurse their

THE PLATYPUS HAS ONE RELATIVE—THE ECHIDNA

young, the platypus mother has glands from which the milk spreads over her skin, and the young animal has to get what it can from this rather casual distribution of its food supply.

The father platypus shows no interest whatever in his wife or children. The mother platypus arranges her nesting burrow to give herself and her young the maximum protection. But if she is attacked in that burrow, her only instinct seems to be to save herself! She does not defend her young at all, but devotes every energy to escape.

The platypus, whether nesting or merely resting, is very timid, and very dependent on having a burrow of some kind to retire to. This, added to its feeding habits, has made it difficult to keep in captivity and practically impossible to transport. Moreover, unlike its dirty and only near relative, the Spiny Ant-eater, or echidna, it is a very clean animal and has to have a constant supply of clean water.

However, there are always ingenious people who attempt the impossible, and usually more or less succeed in doing it.

Mr. Harry Burrell tried keeping the platypus in captivity in order to study their habits in more detail. With a great deal of thought, he constructed an artificial home for the creature—a "platypusary," modeled on the plan of a platypus burrow. After finding, however, that the animals were miserable and failed to thrive in captivity, he abandoned his attempt to keep them, and scrapped the platypusary.

But this device was to be resurrected later on, to play an important part in a famous bit of animal history.

In 1913, Mr. Ellis Stanley Joseph, the internationally known animal dealer, decided to make an effort to bring a live platypus to the New York Zoölogical Park. It was first necessary to devise a suitable carrier for the long sea journey and to make further studies on keeping the animal in captivity. Mr. Joseph and Mr. Burrell went into consultation, and Mr. Burrell resurrected his platypusary.

Three years later, Mr. Joseph, having made extensive observations, and several renovations on the platypusary, decided to attempt the trip. It was a failure.

In 1921, Mr. Joseph again decided to try it.

His first difficulty was that the platypus is a very rare animal even in the one place it lives. It is also an extremely interesting animal because it is the sole survivor of what must have been one of the most ancient of animals—an animal that existed long before man. Realizing its value, the Australian Government has placed very stringent restrictions on the capture of the platypus and on its shipment from the country, dead or alive. This is primarily to stop any possible trade in its skin, for the rather coarse dark brown fur is sufficiently attractive to be of use to furriers.

Mr. Joseph spent a long and harassing time getting special permits for the animals to be taken out of the country, but at last he got these and five male platypus, and installing them in their platypusary he set sail from Australia to the United States.

Before very long one platypus died. Then there were storms at sea and the sides of the platypusary were stove in; their feeding habits were interfered with, and before long only one platypus survived.

Over this lone animal, Mr. Joseph constantly watched. In solitude it occupied that remarkable contraption, the platypusary.

The platypusary is the only thing of the kind ever used in animal transportation. It is about ten feet long, three feet wide and three high; composed of two box-like structures connected by a short, sheet metal tunnel. In one box is a tank of water about eighteen inches deep for the platypus to swim in. Fearing that the animal, gathering too much momentum, would hit himself against the sides of the box, Mr. Joseph placed a metal cylinder in the middle of the tank and the animal swam around this, sometimes emerging to sit on top of it.

Forward of this tank, on a sort of small deck, was another smaller pool of water. This was the platypus' bathroom, flanked on either side by a small sandbank.

The tunnel connected the wet box with the dry one. The dry box represented the platypus' home in the river bank. It consisted of four, small, dry compartments, or burrows, each connected with the other by a small hole surrounded by a rubber washer—just space enough for the platypus to squeeze through. The end

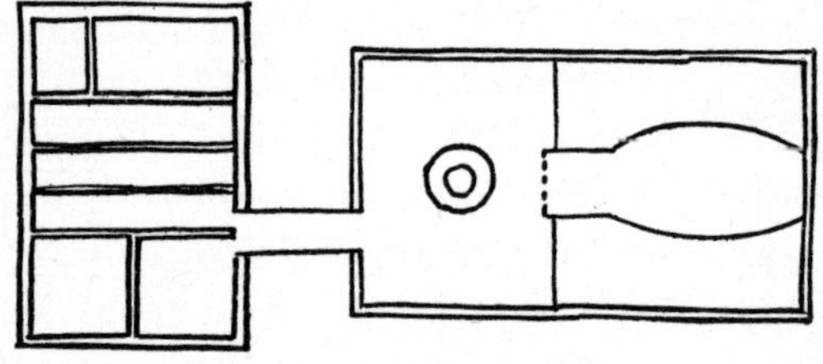

THE "PLATYPUSARY"

compartment of the four was the "nest" and by the time the platypus had squeezed its furry body through the three small openings, enough water had been scraped from its thick fur so that it was quite dry and could curl up and go to sleep without danger of catching cold.

The surviving platypus and Mr. Joseph finally arrived in San Francisco in June, 1922. In San Francisco Mr. Joseph was again confronted by the question of the animal's food supply—this time for the long overland journey to New York. Finally, after great effort and considerable expense, he secured a sufficient amount for the trip.

This was a very hard trip for both Mr. Joseph and the platypus. The small animal was greatly upset by the motion of the train. But at the same time it soon began to be hungry. Whenever the train made a stop, whether day or the middle of the night, Mr. Joseph had to get up to put fresh water and food in the platypus' residence. On July 14th, Mr. Joseph and the platypus, both very much exhausted, arrived in New York and the platypus was taken to the Zoölogical Park.

This was the first and the only platypus to be taken out of Australia alive. It had a short life here, but long enough so that many hundreds of people in the United States have now had the privilege of seeing alive this small strange animal of such ancient lineage.

At first it was very timid and could hardly be touched. Visitors and a crowd around it made it extremely nervous. The Park authorities found they could only let visitors file past the platypus for an hour a day. During this time, however, the platypus behaved very well—or at least, whether from excitement or fear, or from pleasure at an audience, it showed off very well, swimming vigorously around its pond and paddling through the sheet metal tunnel to its nest.

In order to show the animal properly, the top of the platypusary had to be removed, and during the time it was off, the platypus constantly tried to scale the sides of its home and escape. It did not seem to mind the attentions of the keeper who constantly had to lift the small creature off the sides and put it back into its home. In fact, it seemed quite tame, but did not like to keep still in the keeper's hand, even for a few minutes. When picked up, it is very clear that the platypus' skin fits very badly. It is incredibly loosely slung on the body and makes the animal difficult to grip.

But apart from its nervousness and restlessness, the platypus was difficult in the extreme because of its astounding appetite. A meal destined for five normal animals of that size was easily consumed by this one small creature. They may store mud in their cheek pouches, but they are most fussy and demanding about what they put into their ample stomachs. When wild they eat small molluscs, beetles, water insects and small creatures of the river bottom. In captivity they prefer angleworms, very small shrimps, wood grubs of a kind

THE LITTLE ANIMAL IS DIFFICULT TO GRASP

particularly difficult to procure, water insects and oysters. When they tire of one article of diet, which they easily do, they simply will have no more of it. Apparently an animal of only a foot in total length, including the bill and the tail, and only weighing about one pound and a half, can eat its own weight during the course of a day.

The food of the Zoölogical Park platypus in summer averaged in cost from four to five dollars a day. What it would have amounted to in winter was not calculated!

The Zoölogical Park platypus, which had cost about fourteen hundred dollars, untold study, effort and care, died suddenly after forty-nine days at the Park. Everything possible had been done to keep him alive, but it is quite obvious that this animal is entirely unadapted to travel or captivity, however intelligent and careful such captivity may be.

Most of us will only be privileged to see him stuffed in a museum, or if we are very lucky, to see that original skin which so perplexed Mr. Shaw when it was brought to him one hundred and thirty-five years ago in the British Museum, where it now is.

15

THE CAMEL

"I praise Allah for your safe return!" With this time-honored greeting do the Arab desert-dwellers welcome their home-coming friends when a caravan has successfully weathered the strain of a journey across the sands. The salutation is always given in exactly the same words, for the phrase has been repeated and echoed through so many generations that it has become a fixed formula of politeness, rigidly prescribed by Arab etiquette. To omit it would be deliberate insult, a challenge to lasting feud.

News of the arrival of the caravan flies fast, and the neighboring tribesmen flock to the tents of the travelers, eager to pick up the latest desert gossip and to hear the tale of adventure along the route. They come armed, for desert men are always on the watch for lurking danger, but, before entering, they plant their weapons outside, sinking their spears to the shaft in the sand, as a sign that they have nothing to fear within friendly walls. Each guest voices the same greeting as he comes in, the salaam of thanks to Allah for the well-being of his friends. There is good reason for gratitude, for the trek over the waste lands teems with hazard and hardship.

Elsewhere, men skim over the land in rushing trains, cover the seas in speeding ships and even scar the path of the skies in their distance-devouring airplanes. But where the desert spreads, there is only one way to travel

—by camel. The desert is usually pictured as a vast flat of rippling sand, like a sea-beach receding over endless miles. In contour, it resembles any other expanse of land, having long stretches of undulating plain, interrupted by sharp hills and deep-dipping valleys, with every degree of slope and rise that mark the surface of the earth and, for that matter, the floor of the ocean. Nor is the desert uniformly a bed of sand. There are pebbly patches, stony plateaus, rough with rock. High boulders stand in clusters, and whole mountain ranges cut the face of the desert, towering from its gritty base. Furthermore, in cold countries, its barren wastes are buried deep in snow, and its mountain passes are slippery walls of icy crust.

In both hot and cold deserts lying east of the Atlantic, the camel has borne the brunt of travel since prehistoric days. Across the snows of Central Asia, across the sands of Africa and Arabia, the caravan has wound its way, carrying passengers and freight, without interruption for thousands of years. And to-day, in the remote Gobi Plateau, as in the nearer Sahara and lesser deserts, camel trains plod in long files on errands of peace and war, just as they did in the first dim days of civilization. Millions of men still rely on the camel for transport into the far reaches of their untracked country.

The usefulness of the camel is the result of his peculiar body structure. His high build and the perch of his head on the long neck shield him from too close contact with the scorching heat reflected by the sand. His heavy, overhanging brow and thick, long eyelids beat out the glare of blinding sand or the glitter of dazzling snow from his eyes. His wide, slit-like nostrils are equipped with strong, elastic muscles, by means of which he can draw them closed like a pair of shutters against the suffocating blasts of dust in a violent sandstorm.

His feet are well adapted to walking on yielding ground. They are two-toed, the toes equal in size, not encased in hoofs, but protected by a pair of broad nails.

A CAMEL TRAIN

The sole of his foot is a thick pad, covered with a tough outer skin. As the camel steps, the pad expands under the weight of his body, giving him a sure foothold on the shifting sand. On soggy ground, he slips easily and often stumbles to his knees. On rocky routes, the soles of his feet are often sharply cut with deep gashes. Every sensible camel-man carries in his kit a supply of leather patches, and when his camel goes lame, he stops to dress such wounds by resoling the foot with an emergency "shoe," sewed to the edge of the pad.

The cameleer may relieve his animal's foot wounds, and he may cure any festering sores on his back that result from a rubbing load, but if the camel's hump suffers, the problem becomes very serious indeed. In describing a particularly hard journey, the Arabs often use a short saying that gives the whole story in a nutshell. To say, "The camels traveled on their humps," means that their beasts very nearly died. The camel's hump is composed of fat. It serves as a reserve store of nourishment which sustains the animal when there is dearth of food. Other animals, especially those that sleep during the winter months, carry a similar supply of fat which keeps them alive during their hibernation. It is usually spread more or less evenly over their bodies. In the case of the camel, however, the fat is concentrated in the hump. When the animal is in prime condition, the mound on his back is firm and large. When his strength begins to fail, the first sign of weakness shows in the hump. It becomes flabby, shrinks in size, and may even disappear.

Before starting on a journey, camel-drivers examine the state of the hump anxiously. The slightest wobbliness alarms them. It sometimes takes weeks of rest and rich grazing, plus a special diet of dates and grain, to bring the animal back to health. Like horse-traders, camel-men have a bag of tricks for deceiving their customers, and they have ways of bolstering up the hump before presenting the animal for sale. But their buyers are also aware of these methods, and exercise great caution, feeling and prodding the hump very thoroughly before closing the bargain.

Of all the dangers of the desert, the most threatening

is the scarcity of water. Food for both man and beast can be carried, in bales, strapped to the camel's sides. But lugging a sufficient water supply is out of the question. Goatskins filled with water for the men are slung across the camel's back, but the animals must get along as best they can without drinking for days at a stretch. A camel drinks whenever he can, and drinks deep. A normal draught is about eight gallons. If he is really thirsty, it may take twenty gallons to satisfy him. When he is deprived of water too long, he suffers. The first sign of his discomfort is to turn away from his fodder, and when a camel refuses to eat, it is a very bad sign indeed.

Fortunately, his stomach structure is such that he carries a sort of inner reservoir which sustains him through the dry days. The camel is a ruminant, the only one of the cud-chewers, by the way, that has no horns or antlers. His stomach is constructed with three compartments, whereas that of other ruminants has four. The first and second have sections lined with special cells, which become filled with water as the animal drinks. As much as a gallon and a half may be retained in them.

Thanks to the existence of a creature with these many defenses against desert odds, tremendous stretches of territory have been open to trade and travel that would otherwise have remained sealed. The forbidding steppes of Asia, the arid interior of Australia, the pebbly plateaus of Arabia and the powdery plains and hills of northern Africa have been crossed and recrossed by countless camel trains. Our world has been remade from a wilderness of savages and beasts, and we attribute the wonderful change to the courage and invention of great men. Heroes make history, we say. It is no exaggeration to add that some animals make history as well, and of all the creatures that share the earth with man, none, not even the horse, has done more to affect his destiny than the despised camel.

No one knows how or when the domestication of the camel began. It is so ancient an institution that there are nowhere any camels in the wild state, except a few scattered bands at large in eastern Turkestan. Even these are descended from a domestic stock. About two hundred years ago, a terrific sandstorm wrecked the district of Takla Makan, on the lower Gobi. Every living thing in that region is said to have perished in the onslaught of the wind and the choking whorls of sand, except a few camels which broke loose in their panic and somehow managed to struggle beyond the storm center. Their descendants are the only true camels born in the wilds since the beginnings of civilization. The nomad Mongol tribes hovering on the borders of Turkestan used to hunt them for meat and hides, but never captured them.

In ancient times a man's wealth was often measured by the number of his camels. The Old Testament lists the riches of Job as including three thousand camels, and describes him as "the greatest of all the men of the East." To be master of such a troop of transport animals was like owning a railroad, with the added advantage of being able to switch the track in any direction desired.

The eastern kingdoms of antiquity rose to power largely through the use of the camel. Egypt, especially, built her empire on the caravan. Marching single file, as is the manner of caravan formation, the long convoys streamed into the interior, to return laden with gold and unguents and skins and ivory. They brought back also captive black men, taken prisoner, forced to trudge beside the animals, shackled to their gear, to be sold as slaves on the shores of the Nile. Small as she was, the tiny kingdom grew great and powerful, and at one time dominated the western world as it was in the days of Egypt's glory. Her own little realm could never have furnished the stores of wealth on which she founded her might. She drained the nearby provinces of their products, engaging them in trade and acting as the distributing center of export and import. Her treasury swelled with the gold thus gained. She carried war into the outposts of the desert, and exacted tribute from the vanquished with a greedy hand, demanding a heavy toll of their riches and invariably imposing an additional tax in camels, sometimes levying as many as a thousand a year on a single tribe.

In the days of Darius, the Persian king, who built the magnificent city of Persepolis as his capital 2,500 years ago, journeys by camelcade were an everyday affair. Excavations now going on reveal sculptures on his palace walls, showing a procession of two-humped camels, each led by a single rein slung from a wooden peg fixed in its nostril. While the archæologists dig to unearth the ruins of vanished glory, they hear the hum of motors overhead and look up to follow the flight of airplanes in passage from England to India, from France to Indo-China, from the Netherlands to Java. As they level their glances, they catch sight of a slow caravan silhouetted along the horizon. They themselves have troops of camels, essential to their work, standing in clusters near the hangars and garages that house their machines. Like King Darius, they cannot do without the beasts.

Governments of to-day have ambassadors and ministers residing in foreign countries, but in olden times there were no such permanent embassies. Instead, it was the custom for kings to send visiting envoys to far-off courts, as messengers of good will and to encourage trade. Sometimes it took several years to accomplish such a journey. In the second century before Christ, the Emperor Wu Ti, of China, sent a delegation of distinguished men all the way to the shores of the Persian Gulf. They carried an imposing collection of Chinese treasures, bales and bales of presents to be distributed at the courts they visited en route. They came back, laden with gifts for their Emperor, choice examples of the workmanship and products of other lands. Also, they brought back a harvest of new ideas. They had come in contact with the culture of many nations on the long journey, and in Persia they had even brushed against the high civilization of the Greeks, which had become familiar to the Persians during their ancient wars with Hellas. It was a profitable journey for all China—possible only because there was such an animal as the camel to carry them overland.

In war, as well as in peace, the camel has given service. All the armies of old employed him wherever they could. The doughty Kublai Khan made use of camels, when, with his cohorts of wild Tatar tribes, he fought his way to Caraluc, now the city of Pekin, and there established the first empire of the Mongol dynasty in the thirteenth century. Genghis Khan, another bold warrior, and Tamerlane, in his slashing campaigns in Asia, had their war caravans, to carry camp impedimenta and provisions for the soldiery.

There are similar camel units to-day in the forces of the British Army in Egypt. The first one organized was that assembled by the Relief Expedition sent in 1884 to attempt the rescue of General Gordon, trapped among his enemies at Khartoum. It was disbanded at the close of the campaign, but the lack of a camel corps in desert skirmishing was found to be a serious handicap, and a new one was organized, which remains a permanent adjunct to the Egyptian Army. It was kept busy during the World War carrying despatches throughout the Sudan, and is now used in policing the border. It is sometimes called the Frontier Guard.

British Tommies in the Camel Corps never enjoy the early days of their service. The camel's peculiar gait invariably makes them seasick. He advances by moving both right legs in one lurching stride and both left legs in another, so that his rider is jolted from side to side. The heaving motion is as hard to endure as the rolling of a ship in a rough sea. Luckily, camel riders get used to it in time.

For the care of their camels, they rely on natives. In spite of his long history of domestication, the camel is not an easy beast to manage. Very few people have a good word to say for his disposition. He is docile, it is true, after thorough training, but he is considered obedient through stupidity rather than intelligence. He endures the weight of his burdens and the hardship of trying marches under compulsion, but he is very ungracious about it. He is given to fits of ugly temper, and shows his viciousness by snarling and snapping, with occasional kicks and bites. Camels often fight among themselves, and they are not mild about it.

The cameleer is accustomed to their sudden outbursts, and is very careful not to get within reach of their heads, when they show signs of temper. A camel's bite goes deep, and often brings on blood-poisoning. A camel's open mouth is about as pleasant a sight as the gaping jaws of a lion. The upper jaw shows a formidable pair of front teeth, large and powerful. In the adult, only this one pair remains; in the young, the full set, consisting of three pair, appear, but they are shed. The canine, or eye teeth, of the lower jaw are separated from the cutting teeth by quite a broad gap.

Besides biting when he is irritated, the camel has another habit, which is nasty in the extreme. Annoy him, and he spits in your face! When he turns ugly, a swelling develops in the glands of his throat, and he discharges a spurt of saliva at the object nearest him. This disagreeable performance by no means wins favor, and is largely responsible for his bad reputation.

He is especially obstreperous at loading time, and has to be bullied and coaxed into receiving saddle or bales. All during the process, he snarls and shrieks and bellows and roars as though he were in excruciating pain. The drivers pay no attention to his ear-splitting cries, knowing that they are only a protest against assuming the load. They often set up a chanty while adjusting the loads, as sailors do at work, and it is said that the song has a soothing effect on the camel and helps to calm him into a state of submission. Arabs hold that the beasts lose their nervousness when they hear music, and often a cameleer will play on his flute to his sick camel at night. During the desert march, the men strike up a song to cheer their jaded beasts when they lag and loiter, and their pace picks up with the singing.

The expert camel-driver is proud of his skill in the care of his beasts, and knows a hundred and one ways of keeping them in good condition. In northern Africa the word goes that no man is thoroughly expert until he has gone through the grind of the most famous of desert ordeals—the Great Taralum. It is an ancient enterprise, and prevails to this day.

Every year, when the thin crescent of the early fall

moon shows over the rim of the eastern Sahara, a stir and bustle wakes the nomad tribesmen of that region into keen activity. It is the season of the historic Taralum, the huge caravan that annually winds southward to the salt-pits at Bilma. Year in, year out, the long camel train travels the dry route, laden with goods for barter on the journey to Bilma, and returns with packs of salt strapped to the animals' sides. There are times in that region when there is no rainfall for two and three years at a stretch. The trip is made in two stages, with a stop at Fachi for rest and refreshment after the first 300 miles have been covered. On an average, camels travel 25 miles in a day. The Taralum must increase the mileage to 45 in order to bring the camels to water within a week. It is hard going, and man and beast alike are taxed to the utmost.

Added to the dangers of drought and exhaustion and the chance of sickness and death under the parching sun is the peril that lurks behind the hills and in the hidden valleys. At any moment, a band of raiders may swoop down on the caravan, launching a surprise attack with the swift and deadly cruelty of desperado pirates. These desert robbers know no mercy. They know full well that the Taralum carries precious freight. They are eager for the loot in the camel bales, and more eager still for the possession of the camels themselves. They are aware that to deprive a man of his mount in the desert condemns him to a slow, torturing death, but, like the horse-thieves of our sage-country, they are ruthless. Their bands are sometimes two hundred strong; they are fierce fighters, and shoot to kill when after booty.

Hence, it is the custom for travelers going the same route to unite forces against the desert marauders. For the Taralum, they band together in thousands. Traders bound for Bilma come from far and near, assembling at Tabella. They are dark-skinned, sharp-eyed, spirited and hardened desert-dwellers, fierce Tuaregs for the most part, with a sprinkling of Arabs and their black servants. The Arabs are wrapped from head to foot in hooded cloaks of white wool. The Tuaregs wear enveloping robes of darkly striped wool, with scarves attached, which they wind around their faces to keep off the scorching sun.

The time set for the Taralum is in late October, when the few water-pans along the route may hold a little water and the vegetation, such as it is, may be fresh and moisture-soaked. The desert has its own herbage, chiefly thorn-bush, acacia-shrub, and hardy, wiry plants and grasses that thrive in its thirsty soil. They grow in small patches, here and there, and while other animals would starve on their tough fiber, the camel finds them much to his taste.

Every day new arrivals join the camp. It is a time of high festivity, as gay as the Christmas season is with us. The Taralum is the great event of the year, and there is as much commotion and hilarity as at a great fair. Old friends meet and exchange greetings and gossip. Newcomers are initiated into the ways of the Taralum. Those who are making the journey for the first time are known as "sheep"; those who have once undergone its ordeals are dignified with the title of "old traveler." The camp bristles with activity from dawn to dark, and after sundown men meet in the tents to sit around the fire and talk. Tall, copper coffeepots stand in the embers. The evening is passed in story-telling and boasting, in singing, and off in a corner many a bargain is struck.

The entire caravan is subject to the rule of a single

man, appointed as leader. He commands the camel-train as a general commands his army, issuing orders for the morning start, for halts during the day and deciding the route. He maintains discipline with a strict hand, and all obey him without question. The leader of the Taralum is the most respected of men. He has made the trip to Bilma many times. He is wise in the treacherous ways of the desert and her sons, and in the care and control of camels he is expert beyond all others. The members of his caravan rarely address him by name; instead, they accost him by reverential titles, such as "Father of All Camels," or "Knower of the Stars." The caravan takes its direction by the constellations, as do ships at sea, and one who knows the stars is held in high esteem. The Arabs have their own names for the heavenly bodies; what we call the Great Bear and the Little Bear, for instance, they call "The Mother Camel and Her Foal."

To the piercing eyes of the leader, the face of the desert is dotted with guiding signs invisible to others. Though the route has been traveled for centuries, there are no well-marked roads. The sudden, forceful winds quickly erase a trodden path. Where there was a mound of sand but a short time ago, there may be a smooth expanse or even a valley. He who is entrusted with the safe passage across such changing ground must note every tiny landmark. Desert-men often raise a little mound of stones, only a few inches high, here and there on the course to mark the road for their return journey. The leader can spot such milestones from afar. They save the caravan many miles, and prevent the disaster of being lost on the desert. The leader's ranging glance sweeps the horizon. It bores into the sand for scraps left by passing camel-trains and signs of recent happenings. He keeps unswerving watch, for he carries the life of the caravan in his charge.

There was once such a man, a camel-driver named Achmed, who changed the lives of millions of men. He was known as the most skillful desert guide in all Arabia, and every caravan of importance sought him as leader. On his travels he mingled with merchants of many nationalities, and observed their various customs with keen interest. He was most of all impressed by the Jews and Christians with whom he dealt, because their faith was so different from his own. At that time Arabs worshiped trees and stones and other manifestations of nature. Achmed saw that men of other races believed in one God. He thought their faith beautiful and wise, and determined to give his own people a similar religion. He began to preach to the members of his own caravan. They laughed at him. In disgust with their stupid ridicule, he gave up the calling of camel-driver and went into the cities to preach. He changed his name to Mohammed, and made a fortunate marriage with a widow rich enough to help generously in his campaign. His purpose was lofty, and eventually he succeeded in it. He taught his disciples that any one who died fighting for his God would surely enter eternal Paradise, and with this promise of happy immortality in their hearts, his converts followed him into many battles, to spread the religion of Islam. To-day there are two hundred and twenty-five million people in the world who devoutly follow his teachings. Five times a day, every true Moslem turns his face in the direction where Mecca lies, the sacred city of Islam, and prostrates himself in prayer, crying aloud, "There is no God but Allah, and Mohammed is his Prophet."

The first prayer of the day is said at sunrise. In the

desert, the caravan is well on its march by that time, taking advantage of the cool of the early morning, before the sun has scorched the sand. While the sky is still dark, the camel-drivers awake, and by the light of the stars and the glow of the camp-fires, they loose their hobbled camels, which have been tied for the night to prevent their straying away in search of thorn-scrub. The camels are made to kneel, and the business of loading begins.

It takes a full hour to accomplish the loading, for the utmost care must be used in adjusting the bales so that the weight is even on both sides. A badly balanced load means a bad day for both camel and driver, slowing up the animal's pace and ending in a line of festering sores on his back. A camel will carry 300 pounds with ease all day long. If he is very strong, he can manage double that weight, and for a short journey, the burden may be as heavy as 1,200 pounds. However, the camel often makes himself the judge of the amount he will carry, for if he is too heavily loaded, he flatly refuses to rise. Neither coaxing nor cursing will move him, and his driver is obliged to lighten the pack.

No matter what the size of the load is, it is camel nature to protest against it, and this he does vociferously, uttering raucous roars and shrill screams all during the process. Sometimes, they add action to their noise, and jerk the loads clean off their backs, scattering their contents in a storm of dust.

The Taralum numbers easily 7,000 camels, and their clatter and din rends the air for miles. They are of every conceivable color; light brown and dark brown, buff and white, some piebald and brindled, some even of a grayish, greenish or blueish cast. White camels are much admired, but the reddish tan is the rarest and most highly prized. The black camel is an outcast, as Arabs consider him a sure messenger of evil. In their speech and legend, he is the symbol of murder or sudden death.

Camels kneel easily, doubling up like jackknives, with their knees flexed under them, slightly to one side. They rest on their knees during loading, and run no risk of bruising their shins because of the thick, heavy pads, or callosities, that cover them. There are seven of these callosities, one on each knee, one over the upper joint of the hind legs, and a large patch along the base of the curved column of the neck. Otherwise, the camel is covered all over his body with short hair.

Every camel carries a brand, usually burned along the line of his neck or on his cheek, indicating the tribe to which his owner belongs. The *wasm,* or tribal mark, is of very ancient origin. It is of as much significance to the Arab as the coat-of-arms was to the knights of mediæval days, as it tells, briefly, but exactly, his family history. Camels belonging to the Senussi dervishes, the warlike religious brotherhood of northern Africa, display a jagged, zig-zag scar along the length of their necks. It is the name of Allah, as written in Arabic, and is the badge of their masters' fanatic devotion to the cause of the All-Merciful.

A camel in fine condition costs from sixty to a hundred dollars in this part of the world, provided he is full grown and hardened to the march. A baby camel brings only about fifteen dollars, which is rather a high investment, as they are quite useless in their infancy and do not give real service until they are fifteen years old, when they mature. During the first year, they are comically helpless, very wobbly in the leg and altogether dependent on their mothers. When they are six years old, they are taken on their first adventure into the desert, carrying a light load and running alongside their mothers on a short jaunt, so that they may learn by copying their ways. They are still feeble, and if their legs give out, they are picked up and tied on the maternal back, load and all, and carried the rest of the way. It is risky to overtax them before their strength is fully developed, as this may result in crippling them completely, rendering them useless forever. At the age of thirty, a camel is outworked, and he begins to peter out, dying at about fifty. As only one foal is produced at birth, and is a tender little creature at that, the babies get every care. They are nursed by their mothers for a year, though the Arabs are eager to wean them as early as possible, for they want the camel's milk for their own use. It is thick and pleasant to the taste, but it curdles in tea or coffee. The womenfolk pour it into skin bags and let it stand until it forms a rich cheese, or they beat it with a stick into a kind of butter, not much appreciated by the foreign palate, but highly enjoyed by Arabs.

The aristocrat of the desert is the dromedary. No pack-carrier he, but a lean, high-limbed, agile, elegant animal, specially bred for lightness of build and speed. The dromedary differs from the baggage-camel just as a racing steed differs from a dray horse. The finest of them all is the *mahari,* a breed developed by the Tuaregs of the Libyan Desert. The owner of a first-grade dromedary will match him for "points" against all rivals. He is very proud of his mount, and recites the animal's ancestry back through many generations, tracing his descent in a long and ancient line of dromedary lineage.

By an odd and stubborn mistake, the name "dromedary" is generally misused. As a rule, people distinguish the two-humped camel from the single-humped by calling the former the dromedary. The error is almost universal. Keepers in zoos all over the world tell the same story. Parents often stop with their children before the camel cages, and make a special point of emphasizing the difference, almost always pinning the wrong name on the two-humped camel. "Remember, my dear," they say, "the one with the two humps is the dromedary." Time after time, the exasperated keepers take the trouble to correct the statement, but the visitors usually reward them with a polite stare and go away convinced that the zoo attendants are in the wrong. More than once a surprised remark has been heard, such as, "Wouldn't you think the people in charge of these animals would know their correct names?"

The error is of long standing. One of the earliest pictures of the two-humped camel in existence is that painted on the wall of the Chapter House in Westminster Abbey. It is labeled, "The Dromedary," and so the ancient confusion is perpetuated. Great would be the astonishment and high the anger of any Tuareg who might chance to see his lithe and handsome *mahari* so maligned!

Two of the chief sports of the desert result from the swiftness of the riding-camel. Dromedary races were formerly a favorite amusement among Arab chieftains, and they were watched with wild enthusiasm and rivalry that led to high gambling and even to violent feuds. The second sport is hunting the ostrich on dromedary-back, which is said to be the trickiest and most thrilling of desert diversions.

On the fringe of bushland bordering the southern Sahara, the ostrich flourishes, feeding on the plants of the region, such as the cassia, and several gourds and creepers growing as parasites on the low acacia trees. He is a mighty bird, 8 feet tall, 300 pounds in weight, powerful and fleet of foot. He never uses his wings in flight, and keeps them pressed close to his sides when running, spreading them in making a turn very much as the sail of a boat is spread in tacking, so that his balance is not upset by the change of direction. The ostrich is capable

TWO-HUMPED CAMEL

of high speed, especially when alarmed, and can then go at the rate of 30 miles an hour. Hunters have their choice between horses and dromedaries to ride the big bird down, and find themselves on the horns of a dilemma in deciding which to use. The horse is a handicap in that the thud of his hoofs echoes far over the desert, shattering the stillness and giving the alarm, so that the ostrich is away and out of reach even before the hunter sights him. The camel, on the other hand, has an absolutely noiseless tread. The difficulty with him is that there is not a camel alive that can normally be relied on to hold his tongue. Just at the crucial moment, when the hunter prepares to dismount to follow his quarry at close quarters on foot, his camel is likely to let out a roar, sudden and blasting, which puts an end to the hunt then and there. It is possible to train a dromedary to respect a signal for silence, but only if he has been under the control of the same man from his early infancy and is thoroughly accustomed to the ways of that particular master. A camel that has sense enough to remain mute is a very exceptional beast indeed, and even if he can be trusted with his own driver, no stranger can be sure of him. The Bedouin tribes do their ostrich hunting on dromedary-back, using poisoned arrows, but the sport of riding the bird down and lassoing him is a difficult one and its accomplishment is greatly admired.

In countries where the camel is a recent importation, no such use of him in luxurious pursuits is ever dreamed of. In Australia, where the heart of the continent is an immense waterless territory, the single-humped camel serves as the chief means of conveyance. Her own animals have played rather a trick on man by their utter uselessness, except for the fur that a few of them provide. Until the importation of the camel, the greater part of the country was altogether impassable. In the desert mining-camps, at sheep-ranches, at construction posts, and at the lonely, scattered settlements in the interior, he is the one link of communication. About 70 years ago horses were imported into Australia, and now there are a number of automobile roads along the outer ring of the dry region, but for the most part it is the caravan that keeps the shuttle of trade moving. The camels are under the care of Afghan drivers. They haul wool to the trading stations, and come back laden with food and building supplies. They carry wood and coal. They are harnessed to the plow. If a doctor is summoned to an outlying camp, he comes to his patient on camel-back.

The country is so dry that storage tanks of water are a necessity, and this heaviest of burdens is likewise transported by caravan. In some places water is sold by the gallon along the lines of travel, and its scarcity makes a hole in the traveler's purse, for not only he, but his mount, as well, must quench his thirst at any cost. The camel may exist without water for about five days, but after that he shows signs of collapse. There are cases of survival after twelve days, in a temperature of 100 degrees, though very few of the beasts can endure such extremes. It is quite the ordinary thing for a driver to pay five dollars for his animal's refreshment, and one man actually treated his camel to fourteen dollars' worth.

There is a story that has been often repeated concerning the camel's ability to store water within the walls of his stomach. It is a dramatic tale of a desert traveler, who has lost the road, exhausted his water supply, and faces slow death under the parching sun. Even in this extremity, he need not despair. He shoots his camel, extracts the water from the cells of its paunch, and his life is saved! This pathetic story crops up again and again in travelers' tales, although it has never happened. It stands to reason that after several days of storage within the walls of a camel's paunch, the water would be unfit to drink. Even if any one could bring himself to swallow the disgusting liquid, it would only intensify his thirst and doubtless cause horrible suffering, if not death.

Those who are strange to the camel and his ways agree almost to a man that he is the least likable of domesticated animals. His temper, his ugly habit of spitting, his noisy outbursts, his nauseating odor, his sidewise gait that makes riding him an agony of seasickness, all combine to inspire loathing in foreign travelers. One of them has summed it up in a simple phrase, "I never could learn to like a camel." Keepers in zoos say that he is the least responsive to kindness of all animals, and that after years of association he remains entirely indifferent to all friendliness. However, the natives of camel-countries tell another story, as is illustrated in an old Syrian parable:

"Once there were two friends who had each a camel. One of the friends wished to make a journey to Aleppo, so he asked the other if he might borrow his camel.

" 'Yes, Yusef,' he replied, 'if you will promise to treat my camel as you treat your own.'

" 'In the name of Allah, Ibrahim, I promise.'

"So he went to Aleppo, taking the two camels. When he returned, Yusef saw that his camel looked jaded and thin and that his hump had dwindled to almost nothing, while Ibrahim's camel was in excellent condition.

" 'Oh, Ibrahim,' he cried, 'you have not treated my camel as you promised.'

" 'Allah is my witness, Yusef,' said his friend, 'I have done even so. I fed your camel as mine, drove him as mine. He carried a burden no heavier than did mine. Only, Yusef, when I lay down near the beasts at night, as we do to keep off the chill of the wind, I put my head a little closer to my own camel's body.' "

The only familiar evidence offered by a foreign traveler of such sensibility on the part of a camel is contained in a touching story, vouchsafed to be true, of a British explorer who rode a white camel for several months in the Libyan Desert. He says they became friends, and that when the time came for them to part, it was hard for both man and beast. A few hours after he had left the caravan, a runner came to his camp and informed him that the white camel was no more. He was neither sick nor fatigued, but after the departure of his master he had quietly lain down and died. There was no explanation other than that he had died of heartbreak, or whatever it may be that corresponds to it in camel nature.

About twenty-five centuries ago, a journey was made by a wise and seasoned traveler, a native of the Gobi Desert, Fa Hsien by name, over the dreary wastes of his home land, which he thus described: "In this desert are a great multitude of evil spirits and also of hot winds; those who meet with them perish to the last man. Here, there are fair birds above and beasts below. Gazing in every direction as far as the eye can reach to discover the path, one finds no guidance except from the moldering bones of the dead that mark the way."

In recent years, this region has become the focus of interest to those who are engaged in searching out the early history of man and beast. It was in the Gobi that the dinosaur eggs were discovered, which were laid some ninety million years ago. Traces of the existence of men of the Stone Age have been found here, too. At present a large, international group of scientists, composed of Chinese and Western scholars are in camp on the Gobi, engrossed in probing into the secrets hidden under its crust and in its caves. Their findings as to the nature of the country resemble those of Fa Hsien. Fair birds still spread their wings overhead. Hot winds still fan the hard-baked earth in the summer, and cutting blasts sweep across it in the winter. Water is scarce. Mounds of whitened bones still dot the desert, marking the path that other travelers have taken. It is a desolate land, just as it was when Fa Hsien was young.

Such is the country of the Bactrian camel. He derives

his name from the province of Bactria, famous in the time of Alexander the Great for the matchless horses it produced. All the way from the edge of the Caspian Sea eastward to the walls of Pekin, and all the way from the towers of the Himalayas northward to the trans-Siberian Railroad, this camel is the chief means of transport. He serves in an area as broad as the expanse between New York and San Francisco and as long as the reach between Texas and Winnipeg. It holds every variety of land, far-spreading flats, high, snow-covered mountain passes, deep valleys, ledges of slippery rock and salty lake beds. In some parts of the Gobi, the shaggy yak and the sturdy Mongolian pony are employed on short journeys, but where water is scarce, they are of no use.

The Bactrian camel is much shorter than the camel of the hot countries, stockier in build and heavier. He has two humps. His body is covered with a thick coat of hair, which is especially dense along the curving neck and breast and on the upper forelegs. It enables him to withstand the bitter winds and the drifting snows of winter. In the early spring, he sheds most of his hair. The Mongols weave it into blankets and felts, and much of it is exported. The finer hairs are made into artists' brushes.

The two-humped camel is so completely a cold weather animal that it takes the sharp nip of the night air to rouse him to his best efforts. After sundown, his step becomes more brisk, and his stride lengthens. The caravans of Central Asia therefore make their journeys for the most part under the stars. The Mongol is so used to the ways of his beasts and to their swaying gait, that he sleeps as he rides, perched at ease between the two humps. To the foreigner, however, this sort of travel is the last word in discomfort. Unable to bear the intense cold, he sometimes tries curling up in the *joh,* a sort of box, or side-car, strapped to the camel, but usually he finds his torture only increased. The wind whips through the cracks of the *joh;* its wooden floor cramps his muscles; and, worst of all, the rocking, as the camel heaves from right to left, brings on a racking seasickness worthy of the roughest of seas.

At intervals along the well-traveled routes stand caravanserais, crude inns, but blessed by the presence of water-wells. The caravan comes to a halt just before the dawn, and the camels are loosed for cropping, for the sacks of grain and fodder carried for the journey must be made to last as long as possible. Every camel wears a small iron bell around his neck, so that he may be tracked by its tinkling if he wanders away. Nothing ever satisfies his appetite. He eats everything but metal. Such herbage as there is on the Gobi is tough and bitter, but it suits his taste. He thrives on the acrid weeds of the steppes, and drinks the salty water from the stagnant lakes, which other animals could not endure, with apparent relish.

Mongol camel-drivers have to watch their luggage, lest the camels take an occasional nip out of it. Their six-hundred-pound burdens contain bales of wool, furs, grain, or tea-bricks. Sometimes the camels carry whole villages on their backs. The Mongols are nomads, living in rough and ready huts, which they set up on a framework of wood, covered with tautly stretched walls of woven yak-wool or camel's hair. When the pasture is exhausted, or the well runs dry, or when they are afraid of attack by brigands, they hastily dismantle their huts, roll up the felt walls and the wooden props and stack the whole outfit onto camel-back. Man, woman, child and chattels move in this manner from site to site. More often than not, some of the favorite belongings of the household disappear during the migration, for the camels will eat cloth, leather, skins and bones of other animals, and even the wooden utensils or the reed flutes on which the Mongols make their wailing music.

While camels with one hump criss-crossed the hot deserts, and camels with two humps trod the Gobi wastes in never-ending processions throughout the years, in another land camels without any humps at all were serving the needs of man. When the Spaniards entered South America, they found the natives making use of four different animals, quite straight-backed and at first glance more like sheep than anything else. They were so regarded until closer study of their foot and stomach structure and other details proved them of the camel clan.

Until the Spaniards came, the only domesticated animal in service there was the llama, which the Peruvians put to work as pack-carriers, especially in the region of their silver mines. Bolivar reported the use of three hundred thousand of these burden-bearers in the mines of Potosi alone. The Peruvians had an ancient culture, which included an understanding of the arts of mining and smelting. They had a great love of beauty, and one of their favorite decorations was the artificial garden, with branches and blossoms beautifully wrought in gold and silver. The silver ingots were brought down the dangerous inclines of the Andes on llama-back. Since the introduction by the Spaniards of mules and donkeys, the llama has been largely relieved of his duty as burden-bearer, and is now raised in domesticated herds as a meat and wool supply.

The Indians of southern Peru and upper Bolivia take advantage of the existence of another of the humpless camels, the alpaca. It is much like the llama, though somewhat smaller in size, and as its dark brown wool is both warm and waterproof, they make the hides into blankets and ponchos. The alpaca find pasture on the mesas, or flats, in the Andes region, about fifteen thousand feet above sea level, and remain there throughout

ALPACA

all seasons of the year, in spite of the cold and rarefied air. These two, the llama and the alpaca, are now kept in domesticated flocks.

To the south, the wild guanaco and his smaller relative, the vicuña, still supply the Patagonian Indians with meat and hides and wool. Some of the plainsmen of Peru, Ecuador and Bolivia own herds of vicuña and breed them for the sake of their wool, which is extremely fine and soft. It sells at a high price, as the amount from a single animal is small. The natural home of the vicuña is on the windy heights of the mountains close to the zone of perpetual snow. Though his foot has the soft sole of the desert camel, he manages to make his way in the highlands among cliffs and jutting rocks as surely as the firm-footed chamois of the Alpine peaks. The vicuña is extremely shy, quick and nimble enough to escape at any alarm. He is light brown in color, with a cream underbody.

GUANACO

Of the four humpless camels, which are classed together as the Auchenia, in distinction from the "true," or humped camels, the guanaco is the most interesting, as he still exists in a wild state and his habits enable us to guess what those of the true camels may have been before their domestication. In size, he is about the equal of the European red deer. He has a coat of soft, fine hair, pale yellow above and verging into white below. The guanaco range in groups, from half a dozen to a few hundred in number, and, as with most herd animals, now and then a solitary is seen. One of their special abilities is skill and power in swimming, which neither of the "true" camels has in any degree. Their home, in the wild, covers a far range, all the way from the plateaus of the Andes to the Patagonian pampas and even on the islands of Tierra del Fuego, off the west coast, where Darwin watched their capricious behavior with much interest. He found them strangely inconsistent under alarm, sometimes sensing the approach of danger at a distance and making off in a panic, and at other times coming altogether too close to the enemy for their own safety. They showed the most intense curiosity, when the hunters were about, coming nearer and nearer to investigate, apparently as inquisitive about the men as they were about the animals. Darwin had heard that if the hunter lies down on the ground and waves his arms and legs in the air, the guanaco usually cannot resist the impulse to satisfy their silly curiosity, and some of his men tried the trick, with success. Even when the guns went off, they did not rush away. They are very much like sheep in their way of huddling together when they are really scared, and become so confused in a crowded panic that they are easily driven in a mass toward a point where escape is blocked, and after such a round-up their hunters have no difficulty. The Patagonian Indians use this method of hunting them, and get practically all their food and clothing from the animals.

Each of the camels, humped and humpless, has his own adjustments to the kind of country in which he lives, even to a preference in food. While the Auchenia all pasture on grass, the true camels, as though they made a virtue out of a necessity, will turn aside from fresh grass in favor of desert bush or thorn-scrub. How it came about that countries so far apart and so much unlike could all harbor animals of the same kind, was an unsolved puzzle for a long time. The striking differences in size, and the presence and absence of the dorsal hump, divided the camel family into two distinct branches. Still, there remains the resemblance in foot and stomach structure as the clearest evidence that they are related. The guess was that they must all have evolved from the same original stock, and eventually the guess was established as correct.

It was North America that was the original home of the camel. At one time, something like a few hundred thousand years ago, our plains were camel country. In numbers, these animals exceeded all others. With changes in climate and conditions, they began to decrease, and, finally, like the early American horse, they disappeared altogether from our continent. In the far-off time when this continent was connected with North Asia and with South America, bands of wild camels migrated to these lands, and, as they adapted themselves to the conditions in their new homes, they were able to develop and survive as they are to-day.

About twenty-five years ago a fossil camel bed was dug up in western Nebraska. One hundred skeletons were found, lying in positions similar to those of guanacos at rest. It is thought that the early camels were more like these animals than any of the other present-day camels.

Though the camel was once a resident of the United States, he has been found unfit, in his present form, for life in this country. The State of Texas tried to solve the problem of transportation in 1856 by importing a number of camels, and for a short time the humped pack-carriers marched to and from the government freight stations, but the animals did not thrive, and the experiment was abandoned. The camels were turned loose, and they wandered out into the Texan wilds, where they perished. The return of the native, in his later incarnation, ended disastrously.

16

THE WALRUS

MOST animals of the Far North have for ages been the means of life for natives of those barren regions to whom they have meant food, clothing, weapons, utensils, heat, light and transportation. And most of them have been discovered by men from more commercially developed countries with more efficient methods of hunting. This has meant the diminution of the animal; the threatened famine of those natives to whom it meant food and often, as a result, the diminution of the natives as well as the animal.

But ultimately it has also often meant the rescue of the animals and the natives from complete extermination by the advent of more far-sighted outsiders who saw that exterminated natives and animals would be of no value to trade, civilization or anything else. Sometimes the vision has come too late, but in general this has been the history of the reindeer, the caribou, the whale, the seal and the walrus. Of these the walrus is perhaps least important to the outer world, and, in spite of its great bulk, most timid. And for those reasons less well-known than the others.

Judging by ancient accounts, often accompanied by the most fantastic and grotesque pictures of this animal, walrus-hunting as a sport has always been popular, and trade in walrus ivory very profitable.

But in some remote parts of the North, both on Atlantic and Pacific coasts, the animal has played a rôle of utmost importance in the life of those isolated peoples. On one of the islands in the Bering Sea, the Eskimo still start out walrus-hunting when the ice pack comes down. There is quite a ceremony of departure, for hunting this great creature on the floating ice means danger, and the possibility of their not returning is taken so for granted that if the men are gone over twenty-four hours, the community adjusts itself to get along without them, and their wives remarry!

The Eskimo method of walrus-hunting is like a game called "steps." In this game, one person stands about a hundred feet ahead, and back-to, a group, turning around at unexpected intervals. The object of each person in the group behind him is to be the first to tag him without being detected in the act of moving. If you are caught, you have to begin all over again.

The walrus is the person in front. Slowly he emerges from the water, maybe from a hole in the ice where he has been down digging clams with his or her long ivory tusks, and eating this very freshest of supplies. His fat whiskered face comes up first; then the flat flippers plant their rough palms upon the ice and raise the ponderous body after them. The walrus never appears to be in a hurry. He often waits for a wave to come along and help him to beach himself. Then he flops down awkwardly upon the ice or the pebbly shore, full of about a bushel of clams, and prepared to take a comfortable nap.

The Eskimo watch the huge head droop in slumber. Then they advance. The walrus snores. The Eskimo hurriedly conceal themselves behind a snowdrift, for the snore is a sure sign that the animal will wake in a few seconds. It wakes, looks around and snoozes again. The Eskimo creep further forward. The walrus snores. The hunters again conceal themselves, doing their best to keep out of sight, or at least out of scent of the walrus.

This performance goes on until the Eskimo are near

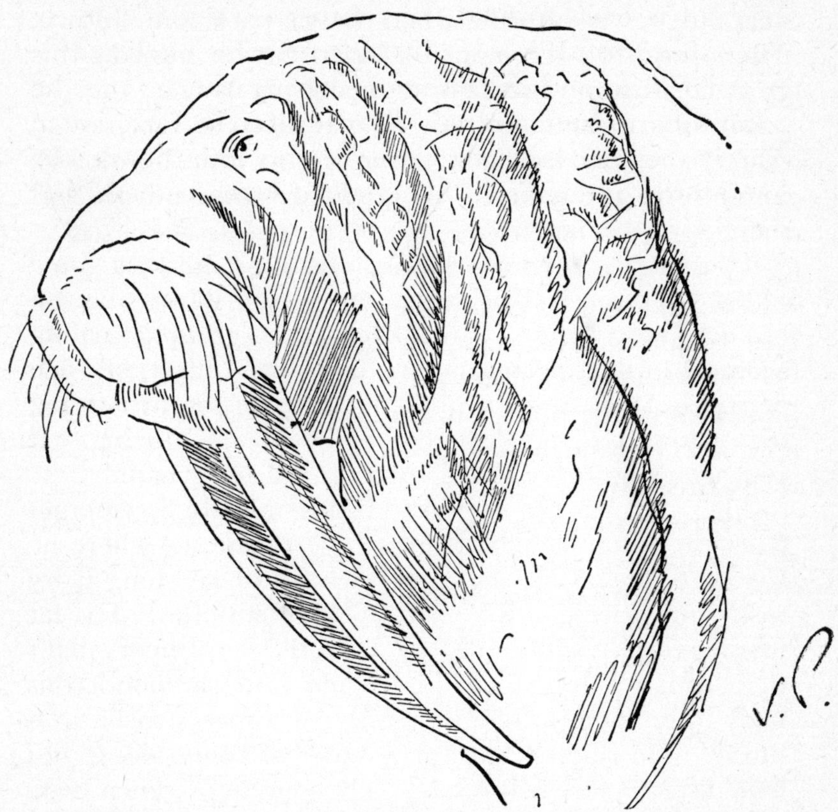

THE WHISKERED-FACE WITH ITS IVORY TUSKS

enough the animal to harpoon him, and another of the walrus tribe is gone forever.

In earlier years, above the booming of water and the crashing of the ice cakes in the "silent North," rose the bellowing of the walrus herds—rose so loudly that it could be heard for miles. But now, although they have not ceased to bellow, the numbers are so diminished that you can travel miles in those same areas without hearing one roar. The destruction of the enormous herds of walrus that once roamed the northern coasts is due to man. The animal's only other enemies are negligible—the polar bear, and certain annoying parasites.

It is safer to hunt and kill walrus when they are asleep either upon the ice floes, the shore, or floating almost vertically in the water. An angry walrus can do a lot of damage with his flippers as well as with his long tusks, and if one of a herd is attacked, the rest will join in in a well-concerted revenge upon the attacker. If a small boat advances upon a lone walrus, the walrus is apt to rear up suddenly and grasp the end of the boat with his flippers, and the hunter in the boat is then apt to abandon ideas of hunting and get a taste of what it feels like to be hunted!

The reason for the disappearance is over-hunting, and this in spite of the fact that the animal on both coasts lives in a part of the North only accessible for a few months of the year, because of ice. The Atlantic walrus lives off the islands north of America, eastward to about the mouth of the Yenisei, and in Hudson Bay, Labrador and Greenland. The Pacific walrus, a little the larger of the two species, lives along the Bering Strait, Kamchatka, Alaska and the Pribilof Islands. But in spite of the inaccessibility of its range, it has been valuable enough commercially in the past to pay men to equip themselves and to come to the North to hunt it. Recently the difficulties and its diminished numbers hold out hope that hunting will become less and less frequent and that the walrus may get a chance to increase. Only, in all probability, to be again diminished because of the ivory of its tusks and because it provides good sport for adventurous hunters.

This brown-furred animal is a member of the seal tribe, and in general looks like the seals. Both seals and walrus have flippers adapted for life in the water, five digits on each foot, and the tip of the foot webbed. Both are able to turn the fan-shaped hind flippers forward for land progression. The front legs of both are enclosed in the loose skin of the body down to the elbow. But the walrus is a far bulkier, more awkward creature than is the sleek, graceful seal. It has a shorter, thicker neck and immensely powerful shoulders. This neck and shoulder region contain enormous masses of blubber, highly valued by the natives of these regions. The eyes are small, but slightly popping; the ear openings are practically hidden in the folds of loose skin. Nor has the seal the long tusks which represent the canine teeth of the walrus and are fixed surprisingly loosely in their bony sockets in the upper jaw. Both the fur and the oil of the walrus are far inferior to those of the seal. In fact, walrus fur is not valuable at all, and often hardly present!

Young animals have lifeless brown fur, but as the animal grows, the fur disappears, until the adult walrus is practically bald all over, especially the old male—and a most disagreeable sight.

About eleven feet long, almost as wide as long, and weighing some two thousand pounds, this animal dives and swims with ease, swimming under water and rising to the surface to puff and blow and take in air. But on the shore he waddles with difficulty on his flat, rough-palmed flippers, his bulging body falling in folds and wrinkles about his feet. Adding greatly to the unfortunate appearance of the walrus are the warts, wrinkles, and bubbly excrescences on its bare hide. Its skin is so loose and baggy that it appears to have no muscles. The flippers, although in reality strong and flexible, look flabby and helpless. But in spite of his figure, there is something very massive and impressive about the pompous, whiskered face with its long ivory tusks. And in spite of the fact that the huge animal is not beautiful, he is harmless and unaggressive unless attacked.

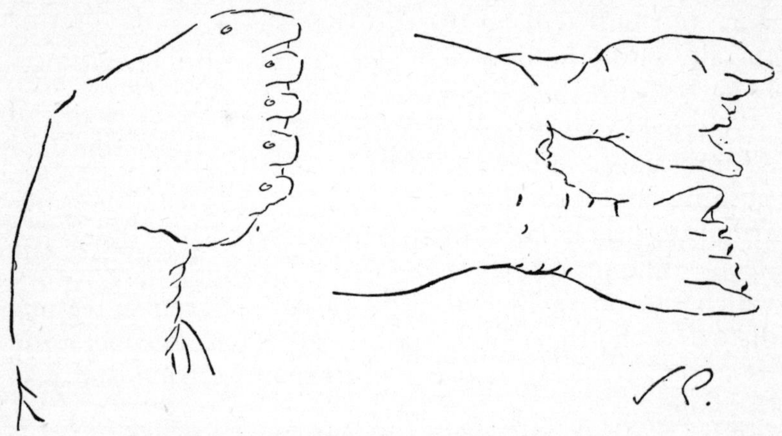

FORE AND HIND FLIPPERS OF WALRUS

The walrus apparently has a very loyal nature, both to its fellow walrus and to its young. The female will defend her young to the death. Hunters tell us that she will roll them off into the safer regions of the water where she holds them in her flippers, and finally, if unsuccessful in guarding them, will attack the hunter with ferocity.

The young are born on shore in spring or early summer, one, or at most two, at a time. But otherwise, the walrus is not often met with either on shore or in the open sea. It usually lives on the ice floes near the coast line. This is because its food consists of clams and crustaceans which it finds in the waters of these regions and digs up from the bottom with its tusks. The shells are crushed with its grinding teeth.

On the thick, loose ice near the coasts great herds of walrus move around, or sun themselves, resting on each other like huge pigs, and grunting like pigs as they heavily shift their position.

"Occasionally," writes one observer, "in their lazy phlegmatic adjusting and crowding, the posteriors of some old bull will be lifted up, and remain elevated in the air, while the passive owner sleeps with its head, perhaps, beneath the pudgy form of its neighbor!"

A crowd of walrus getting up on the ice together is a funny sight. They crowd each other to the last degree, pushing and prodding with their long tusks, lopping around on their fat bodies, gruntingly settling down for sleep. The proddings of the thirty-inch-long ivory tusks, which can do such damage to boats, apparently do not disturb each other. The young walrus has no tusks until its second year, but it gets its full share of prodding just the same.

The ivory of these tusks, which on the Pacific walrus curve slightly more inward than on the Atlantic, has a certain value and is used for the coarser uses to which ivory is put—sword grips, cane and umbrella tops. In earlier years it was also used for ornamental carving. In spite of the length of the tusk, only a few inches of it are usually of much value commercially. A coarse, yellowish, bony pith occupies three-quarters of most walrus tusks. Also, many walrus tusks are broken and chipped.

The hide is very useful indeed to the native, for harness, ship rigging and anything else for which such thick, tough material is useful. With infinite patience the St. Lawrence natives pare down and prepare this hide to make their *bidarrahs,* or skin boats, sewing it with sinews to the wooden, whalebone-lashed frame, and then oiling it. When it is finished, the boat has an almost transparent appearance but is very strong and light.

Probably scarcity of other food has taught the natives not to be choosey, for they seem to enjoy walrus meat, which to our minds is not very delicious. It is acrid and tastes overwhelmingly strongly of the molluscs upon which the animal feeds. But disagreeable as it may taste to us, it is of utmost importance to the native.

Trade in skins and ivory was carried on in a small way, we may assume, for centuries, but in more recent years this trade was taken up in earnest and so diminished the supply of walrus, that, together with over-whaling, it diminished the supply of Eskimo as well—almost to the vanishing point. It cut off one of their main sources of food, heat, and light. Unfortunately, in Alaska, overshooting of the native caribou, over-whaling, and overhunting of the walrus all occurred at about the same epoch. At this same time, the crews of the trading and hunting boats introduced the natives to whiskey, and the natives, facing famine, took to the warming drink over-enthusiastically.

Living where they were so difficult to reach, in a way that outsiders so little understood, speaking a language so few other people could understand, it seemed at one time that this race was doomed. But the very factors leading to this condition brought their plight to the eyes and ears of intelligent men who introduced certain relief measures, chiefly the introduction of the domestic Siberian reindeer, which proved successful beyond imagination, and brought into these inaccessible regions an animal more useful to the native than those whose depletion had caused its introduction.

17

THE REINDEER

Two rich Venetian merchants named Polo sailed out of the Adriatic many years ago on a voyage to the East. While the two merchants were trading with the ruler of those far-away lands, a son was born to one of them back in Venice. When the father came home again, this boy was already fifteen years old. The father was so delighted with his son that he decided to take him along on the next trip. Very shortly the ship again sailed out of the lovely, water-surrounded city of Venice, not to return until the little boy was a young man—one of the earliest explorers, Marco Polo.

In those days there were no great steamers covering three thousand miles of ocean in four days. It took months for the sailing vessels to reach another port, and land traveling was done by horse or camel. To cross one country often meant a two months' journey, and in the heavy winters of the North, people could not travel at all, but had to wait somewhere for a warmer season to come.

When merchants set out on a trading expedition to another country, they expected to spend a long time there. They took with them, not order blanks, but all the actual goods they intended to trade with, and gradually, on their way from place to place, they exchanged these goods again and again until they had acquired the most valuable and most easily transported exchanges possible.

Marco Polo's father was a prominent citizen of Venice and on his previous trip to the East he had been able through various connections to make a very powerful friend—the most important ruler in all the East, Kublai Khan, the Great Khan of the Mongols and Emperor of China.

To Kublai Khan's magnificent palace young Marco and his father planned to return. They sailed to Constantinople, through the Bosphorus and up the Black Sea to the Crimea; then by land they went east to Bokhara, on through Persia to Afghanistan; from the high plateau of Pamir across the Gobi Desert, and finally they came to the court of the Khan, in the city now known as Peiping or Pekin.

In the Khan's domains they spent many years, sometimes at his luxurious court and sometimes wandering among the deserts and mountains of Central Asia. Marco even became ruler for a short period over a small domain of his own. The Khan was charmed with his visitors from a distant land and would gladly have kept them in his country forever, but the time came when the elder Polo felt they must return to Venice, and so, after this absence of many years, they came back to their native city.

To the Polo family, the journey had been a very profitable one. Before the astonished eyes of their relatives and friends, they slit the hems of their shabby traveling garments, and there poured forth a glittering pile of precious stones—showers of rubies, emeralds, sapphires and pearls!

But to the world, the journey was even more profitable,

for it gave people for all time to come an accurate record made by one of the earliest explorers in the East.

Almost immediately after the Polos' return, Genoa and Venice went to war at sea. The whole Venetian fleet was captured and among the prisoners was Marco Polo. While he was in prison in Genoa, he wrote a famous account of his journey.

His stories were not very widely believed. People thought he had grossly exaggerated what he had seen, or that he had made up the whole thing. It has taken centuries for the truth of his observations to be proved. But one by one the statements that seemed so fantastic have proved to be exact and unadorned truth.

One of Marco Polo's tales ran that in going out from the citadel of Karakorum and north from the Altai Mountains for about forty days, the traveler reaches a country where people make their home in the forests and feed on the forest beasts, especially the deer. These people, writes Marco Polo, understand how to tame deer so that they can be both ridden and driven, like horses or donkeys!

No one believed this story. They could not imagine riding or driving a deer. It was put down as just another fantastic story by the highly imaginative Marco Polo, whose name came to be used as a synonym for "liar."

But although the geography is a little confused, due to changes of names down through the centuries between then and now, Marco told the truth about these forest animals and their use. The animal that was ridden and driven in that Asiatic north was the reindeer. For hundreds of years this animal has been used in this very way by northern peoples,—the Samoyedes of the Siberian tundra, the Lapps and the Scandinavians.

Four centuries later, the same tale was told, and not

much more widely believed, by another trader, a Portuguese who had been in Lapland. He went into further detail about the strange animal that the Lapps used as a domestic beast. He wrote that it was about the size and color of a donkey, had split hoofs and the head and horns of a deer. He added that the horns were covered with a "sort of wool." Moreover, he noted that whenever the animal moved, its feet made a clicking noise.

This is a short but accurate account of the animal known in Europe as the reindeer and in America as the caribou.

Perhaps it might have been simpler and nearer the mark if the Portuguese trader had merely said that the animal he saw was a large deer, but of course a donkey was more familiar to his southern eyes, and the description is accurate as far as it goes.

Reindeer differ in size in various localities, from that of a large deer to almost as small as a goat. Typical reindeer stags stand some four feet at the shoulder and weigh about six hundred pounds; the doe is slightly smaller.

The American Indian name for the reindeer was a word that sounded like caribou, and this name has been adopted by the white man. The word "reindeer" is derived from Germanic words meaning "to run" and "animal." So it has happened that our native reindeer, living in North America, Arctic Canada and Alaska, are known by the old, Indian-derived name, "caribou," and the same animal in Europe is known by the name "reindeer." Between the American and the European forms, there are some differences of color and of size, but in general the animal is large, densely furred, migratory, antlered in both sexes, and bearing the peculiar foot structure known as "cloots" or "dew-claws."

The various ways of designating these animals are confusing unless you use their scientific names. If you know exactly where your animal comes from, you may perhaps safely speak of it as a Barren Ground caribou, or a Woodland caribou, but if not, you are on rather unsafe ground. The Barren Ground caribou takes its name from the stretch of desolate country from Hudson Bay to Great

THE CARIBOU

Slave Lake. The European reindeer comes under this general division, which includes in its American range Greenland, Ellesmere Land, the Arctic Islands and Alaska. This animal generally migrates at the cooler seasons of the year to the woodland areas farther south. The Woodland caribou ranges from Maine to Canada on the east, and from Idaho over the Rockies and to British Columbia on the west. In this division is the largest and handsomest of all caribou, the Osborn caribou of the Cassiar Mountains of British Columbia. This

Woodland caribou is apt to move further north, to the mountains above timber line in the warm seasons.

But whether the animal spends most of its life in the lower forested areas, or on the tundra, those bleak stretches of land with almost permanent substrata of ice, the scientific name remains *Rangifer,* and the reindeer of Europe and the caribou of North America are both reindeer.

Whatever their particular variety and whatever their home, reindeer are born in the North. We may have our birthdays in any month of the year, but the reindeer's or caribou's birthday always comes in late spring or summer. This is chiefly because their whole behavior and residence are governed by an effort to get food in the changing weather conditions of their home in the North. Wherever it is coolest and wherever the does can best avoid the plague of torturing insect pests—the gadflies, and those horrible, ever-present mosquitoes of the Far North—there they go to bear their young. They often choose the windswept and even snow-covered tundra, which would seem to a human being a particularly unprotected spot.

The soft, reddish-brown young are not able to take care of themselves immediately after birth, although they can walk almost at once. They are dependent on their mother for food. The doe nurses her fawn for about two months, after which they begin to feed for themselves on the favorite food of their kind—mosses and lichens, especially a lichen known as reindeer moss of which huge areas grow on the tundra. The young stick very closely to the doe and the doe shows little interest in anything but her fawn. Sometimes there are twin fawns, but usually only one. The does, however, occasionally seem dissatisfied with only one, and show signs of wishing to adopt another. Two does sometimes fight for the possession of an adopted child for whom they care as protectively as they do for their own.

Reindeer begin to grow antlers sooner in life than any other deer. Even at birth there are two bumps on the forehead of the fawns. These are the beginnings of the imposing antlers they are to carry later in life. In a few weeks these bumps are well-defined knobs and in about six weeks they are some eight inches long.

In the summer season the animals are all more or less solitary, stags and does alike. They move about in twos and threes and usually stay in the woods in the daytime, coming out in the cooler evening to feed. Summer is a very hard season for them as to temperature, for although their summer coat is thinner, finer and smoother than their winter coat, it is still pretty thick and warm, and they feel the heat badly. When on the run, they sweat as dogs do, with their tongues hanging out of their mouths, and often take a mouthful of snow, apparently to cool themselves. In spite of this heavy coat, they are tortured by the blood-sucking flies and the swarming mosquitoes of the northern forests and swamps.

The reindeer's fur, like other details of their appearance, differs somewhat, depending on what portion of the world they live in. In general, the farther north the animal lives, the smaller it is, the shorter its antlers and the lighter color its fur. Domesticated animals too are generally smaller and have less imposing antlers. Recently, however, the owners of the large Alaskan herds have begun to breed their Siberian-imported animals with the native caribou, producing a larger, stronger race thereby. The Mountain caribou of northern North America is a large, dark animal with immense antlers, but the Arctic or Little caribou of the tundra of the far north of Europe and America is a much smaller, paler animal, with thinner antlers. In Spitzbergen there is a very small reindeer, almost as small as a goat.

The typical reindeer fur is rather mouse gray or grayish-tan, lighter on the sides, neck, rump and under the body. There is usually a white ring around the large, gentle eyes, the small ears are light, and often there is white around the leg slightly above the hoof. Farther north, the reindeer approach pure white in their pelage, especially in winter.

At the beginning of the summer neither the stag nor the doe has antlers, but before the end of that season the stags have developed fairly good ones. The doe's antlers develop more slowly and are never very large, but the reindeer is the only deer in which the does have any antlers at all.

At this time, and all the time the antlers are developing, they are covered with the "sort of wool" described by the early Portuguese trader. We call this the "velvet." The velvet, although it is furry on the outer surface, is not fur. It is a kind of skin and has the same blood vessels and nerves that skin has. Consequently, while the antlers are covered with it, they are very sensitive and of no use to the reindeer.

When the antler is full grown in the autumn, this velvet, having performed its duty of providing the necessary moisture to the growing antler, begins to dry up and peel off. It does not fall off without help. The animal gets rid of part of it by rubbing his antlers against the trunks of trees. If some of it still sticks on obstinately, he scrapes it off very carefully with the sharp edge of his hind hoofs.

When all the velvet is off, the antlers emerge in their burnished glory, a rich reddish brown, almost orange; sometimes almost crimson. They have taken about five months to develop and now when they are so handsome and large, they almost at once begin to dry! Before long they fall off, and the ornament that took almost half the year to reach completion is gone!

The reindeer have the most handsome antlers of all the deer tribe. Varying in the different reindeer from long thin branches, to stocky wide ones, they strike their most impressive state in the many-branched, palmated antlers that rise high above the head of the Osborn caribou. The high branches with their broad leaf-like palms, rise up on sturdy stems, curving gracefully backward from each side of the forehead, before the ears. In front of them, facing in the other direction—forward over the nose—is the massive, palmated brow antler. A herd of reindeer on the move seems like an advancing forest of burnished, towering antlers.

From the time that the stag was royal game in Europe, there has come down to us a special vocabulary, once used in the chase and now in general use, to describe various parts of antlers. Hunters are very keen on capturing beasts whose antlers have a great number of "points." This means any knob or jut on the antler. The usual number of points in a buck with good antlers is twenty-eight to thirty, but they are known to achieve as many as forty.

The antler above the brow tine is the "bez tine"; the next one above is the "trez tine"; the fourth branch and the rest of the structure above it form the "crown" or "sur-royals." Often the antlers of individuals in the same herd are quite different looking, and sometimes the antlers of one animal are not symmetrical. In general, antlers, like trees, add branches every year until the buck is fully developed. They then add no more branches and the branches seem to decrease in number and size.

Antlers, in contrast to the true horn of a cow, are not, strictly speaking, horn. They are bony in nature and not hollow. The antlers of the reindeer are rather springy and for that reason all the more easily interlock if two bucks get into a fight, as they often do. It is amazing that they do not also get entangled with the dense growth of the forests.

With these shining antlers, the heavy pale winter pelage and their dark polished hoofs, they are the living counterpart of those Santa Claus reindeer that go prancing over chimney tops delivering presents,—although the famous poem on the subject speaks of the "tiny reindeer," and reindeer certainly are far from tiny. Santa Claus sits in a well-balanced sleigh, and the reindeer of those pictures usually appear to be harnessed to that sleigh very much as a horse is harnessed to a carriage.

In the remote North this is far from being the case. The reindeer is used instead of a horse in many parts of the world, especially in Norway, Siberia and Lapland, and, more recently, in Alaska. It is fond of people and makes a tractable domestic beast. Some peoples of the far north of Europe, in regions where horses do not thrive, are very dependent indeed upon this animal for transportation, for food and even for clothing. Unfortunately, in some of those same places, there is another animal, not human, also very dependent upon the reindeer for food—the wolf. A hungry pack of wolves will follow a herd of reindeer or even a single animal for miles. Desperate from starvation, these cruel animals will even go into villages and capture domestic reindeer. And only a little less ruthless in its pursuit of this very gentle animal, is the gulogulo, or wolverine, a smallish animal with the evil preying habits of the wolf. In regions where there are wolves, the owner of a herd of reindeer faces a very difficult problem.

For you cannot shut up and protect the reindeer in a stable. Although a very obedient and willing worker in working hours, the reindeer must be allowed to live its own life out of working hours, and this very necessity for freedom exposes it to the attacks of enemies. When its services are required, it is lassoed, although some tribes have domesticated their animals to a state where they will come at a call. This is particularly true of the sturdy reindeer used by the Soyots of northwestern Mongolia as a pack animal. The Samoyedes ride the reindeer, but it is much more frequently used as a driving animal. The life of these Samoyedes is so bound up with that of the reindeer that they have given to several of the months of the year names connected with the calving of the animal.

Not only is the reindeer independent as to stabling, but it also cannot be driven in the firm and rather heavy harness used to control the activities of the driving horse.

In Alaska where the importation and domestication of the Siberian reindeer has probably saved a population from starving, the harness has become more sophisticated and more secure. There are light but firm traces, and breeching, and the sled is flat-bottomed and much safer for the occupant. But in regions more remote from contact with the rest of the world, to drive a reindeer is quite unlike sitting in a sleigh behind a horse. The horse is very firmly attached to the vehicle behind it. The bridle and bit, collar, traces and breeching all serve to keep the animal bound to the carriage and the driver. And above all the bit. The driver of a horse guides and controls it by two reins which run from the metal bit in its sensitive mouth, pulling the reins to right or left to indicate to the horse which way to turn. Every part of the harness fits together to keep it on the horse and to keep the horse attached to the carriage and at the disposal of the driver.

Not so with the reindeer in the Far North! The reindeer will not tolerate a bit in its mouth, nor will it stand the heavy collar, the tight girth and the confining traces and breeching. It cannot bear to have its head jerked or pulled. It is collared, but with a soft hide collar on which the fur is left. To this collar in the lower center in front of the forelegs is attached a single line or trace, running under the animal's body between the legs, but not touching it, and fastened at the other end under the prow of the canoe-shaped sledge in which the passenger sits. This sledge is keeled and light, more practical than a runnered sledge because it does not cut the snow. In it sits the fur-clad, tightly packed-in traveler.

Balancing this snow boat is a perilous adventure to the novice. Reindeer are not inclined to go slowly while the driver is learning to balance so that he will not be hurled out of his vehicle. A single rein is fastened to the base of one of the antlers, and to this rein at the other end is attached the driver. The driver of a horse holds two reins tightly or loosely as the occasion demands. At any moment he may drop them if he wishes to. The driver of a reindeer carries one rein only and that rein is bound tightly to his right wrist. This is because the reindeer has more control over his driver than the driver has over the reindeer. If the pulka overturns, as it often does with the inexperienced driver, the reindeer will not stop at a shouted "whoa"! It is very important that the reindeer and the driver shall not be separated in the vast snow fields of the North, and the driver left without any way of getting through the deep snow. So by binding the rein to his wrist, the driver is at least sure of accompanying his steed, although for several hundred yards he may do this by being dragged by the animal over the snow and ice in a most uncomfortable manner, until he succeeds in getting hold of his sleigh, righting it, and getting himself inside it again! The rein does not help him much in controlling the speed of his swift animal, as it must be held very loosely. Attached to the base of the back antler, it festoons down beside the animal so loosely that it nearly touches the snow. To guide the animal to the left or right, it is thrown over the animal's back in the desired direction.

In going down hill, the sleigh is apt to overtake the reindeer unless some kind of a brake is used. The driver has his choice of getting out and walking and holding back the sleigh, of using a pole if he is expert enough to manage it, or of sitting back-to his reindeer and using his own legs and heels as brakes! It all sounds very uncomfortable, but after one has become expert in reindeer driving it is very exciting to be carried over the snow by this swift sure-footed animal.

Training reindeer for domestic use is highly skilled work, and they are good for only about four years of service after all the training has been done. From three years old to about six they are in training; at six to eight years they are at their best in service, and then they begin to get old, and at nine their best time is past. Other species of deer have been tamed, but no other deer has been domesticated. One of the greatest advantages of reindeer as a driving animal is that unlike the sledge dogs of the North, it appears to have no objection whatever to the wind, whether it comes in its face or not. And of course the maintenance of reindeer is infinitely less costly than that of a dog for which food must be provided and carried along on trips. In fact, in case of extreme need, the reindeer can even be converted into very palatable food for its owner. It must be extreme need, because the owner of reindeer is generally fond of these gentle, independent animals who are such good workers and cause him so little trouble to care for. The Lapps are so proud of their reindeer herds that they have written many songs consisting entirely of praises of their handsome herds.

Unlike its European relative, the caribou of North America has never yielded to domestication. The European animal will not stand being put into barns or stalls; it sticks around close to the settlement or village without being stabled or tethered, finding its own food and sleeping quarters. But the caribou will not stand captivity at all, and is not used in harness.

In Alaska, however, you will see reindeer in harness being used for the transportation of both people and goods. These are not native reindeer. They came to Alaska from Siberia as emigrants, under escort of the United States Government, and have proved to be great successes as citizens.

Some forty years ago, the Eskimo of the northern coast of Alaska were in danger of starving. Whaling and walrus-hunting had been carried on excessively off that coast, and the native caribou were disappearing from inland because improved firearms had made their shooting easier. Food was getting scarcer and scarcer. With no railroads or steamers to connect them regularly with the outer world, and their native food supply rapidly disappearing, the natives faced famine.

Across the straits from this starving country, the nomad Siberian tribes derived secure support from their large herds of domestic reindeer. For years, the Alaskan Eskimo had been aware of this. In native boats they had crossed the cold waters separating Alaska from Siberia to trade oil with the Siberian deermen for reindeer skins. But the small, lightly built native boats precluded all thought of transporting the animals themselves, even if the idea had occurred to any Eskimo.

In 1885, Dr. Charles H. Townsend, then a young scientist cruising on a revenue steamer in the Arctic Ocean, noted this condition and in the report he made to Washington on his return, he suggested sending boats up to Alaska to introduce the Siberian reindeer into that country. Alaska has an enormous unforested area suitable for reindeer to graze upon—land covered with reindeer moss—and all the reindeer needs is a good supply of food, for it is accustomed to forage for itself.

Agents were eventually sent up to Alaska to investigate and to make arrangements with the Siberian reindeer owners. They at once encountered difficulties. The natives on the Siberian side were nomadic people, superstitious, knowing nothing of the value of money and speaking a practically unknown language. However, in 1891, proceedings began in earnest. The Russian Government coöperated by instructing their officers on the Siberian coast to give what help they could. The United States agents took with them articles with which to pay for the animals—useful articles such as tobacco, oil, utensils of various sorts, in return for which they procured live reindeer. Here they encountered superstition. The Siberian deermen were extremely superstitious about these animals who played such an important part in their daily lives. If any accident happened to a member of a family who had sold a reindeer, the mishap was laid to the sale and the neighbors immediately became very loath to part with their animals. Utmost tact was necessary to deal with these wild people at all.

The Eskimo on the Alaskan side knew nothing of the training or care of the domestic animal. Neither did the government agents. Therefore, wild as they were, Siberian herders had to be taken along to Alaska. Nomadic and dissatisfied, they proved no very great addition to the community, but for three or four years they were the best that could be had. After this time it became obvious that the experiment was going to be a great success and in 1894 the first colony of Lapps ever brought to the United States were landed at the Teller Reindeer Station, after a journey of over twelve thousand five hundred miles—sixteen people in all—seven men, their wives and children. These seven men were to undertake in earnest the instruction of the Eskimo in the training of the Siberian reindeer to whom they were so accustomed in their own northern home.

Reindeer herds if properly cared for double in size every three years. Under the intelligent, experienced Lapps, the Eskimo very soon became expert in their use and care; the animals throve and more were brought from Siberia.

The importation of these reindeer opened up a new life for Alaska—communication with formerly inaccessible places, communication with the recently discovered goldfields, communication with trade and the outside world. And in addition, it provided transportation, food and clothing and healthy occupation for the Eskimo. At the present time there are thousands of Siberian reindeer in Alaska. Many of the huge herds, like the Lomen herd, have become world famous. These herds, when they are not at work, still roam free, but carefully watched by herders. A great annual event of Alaskan life is the reindeer fair held by the Eskimo who bring their animals from near and far to exhibit them and the sledges and harness and to enter them in the sledge races. The beautiful, strong, white-furred animal standing in harness in the middle of a group of admiring fur-clad Eskimo, its powerful antlers spreading into the air above their heads, is to our more southern eyes a sight like something out of a stage setting.

In parts of the world where there are no cows, this useful animal is also utilized as a milk supply, although the amount of milk from a reindeer cow is very scant compared with the amount from one of our cows. In the more desolate parts of such countries, however, the natives boil reindeer moss in reindeer milk as almost their sole article of diet.

Very long ago, in some age of the world which we only know about through the discoveries of science, there was a time when reindeer were apparently the most plentiful and most important animal on the earth. This was known as the Reindeer Age. The reindeer must have formed a very favorite subject for drawing in that far-off time, for in ancient caves there are still to be seen crude stone carvings and drawings perfectly recognizable as this animal. Although the reindeer has disappeared from many parts of the world where its records and the extent of its fossilized remains show it to have been of such importance—the largest and most plentiful of the large mammals of its time—to dwellers in the bitter snow-swept countries of the North, it is still among their most important possessions. Possibly because of its enemy, the wolf, the reindeer is usually very much afraid of any kind of dog. Sometimes dogs cannot be kept at all if one keeps reindeer. Sometimes, however, it is possible to use dogs to keep the herd together, as sheep dogs are used with flocks of sheep.

But if it is afraid of wolves and dogs, it seems to have

a natural friendliness toward man, and unless it has learned to fear human beings through being hunted, its curiosity about them is remarkable. Even when a hunter has killed one or two of a small herd, the other animals of the herd, instead of fleeing, have been known to come up and nose at the hunter. It often becomes a great pet with its owners and the children around the settlements. Like many other wild animals, especially vegetarian animals, these reindeer are very fond of salt and can be lured by it to stay near the dwellings of their masters. In traveling with reindeer, if not watched, they will nose out the salt sacks on the sledges and lick up all the contents.

Caribou although not domesticated are, like their relatives, providers of very fine warm winter clothing for the people who live among the biting cold and blasting winter winds of the North. Their meat is also good to eat.

In fact, reindeer are one of the most economical and useful of animals. They provide fur and skins for clothing; their meat is highly edible. They are good steeds for riding or driving, and where there are no cows they can give a small quantity of drinkable milk. They cost very little to keep in captivity for they do not wish artificial shelter and prefer to eat the mosses and lichens which they find for themselves.

They are not only well able to take care of themselves when they are in man-owned herds, but when they are roaming around in their wild state they are one of the best equipped animals known. They swim as well as they walk and as fast if not faster. Their senses are keen and alert. Unless the wind is away from them, their nostrils scent danger incredibly swiftly and surely. Their feet are constructed to manage bogs, barren ground, woodland tangles, snow and ice. Their fur is thick in winter and thinner in summer, and at any season acts as a splendid protection against rain and snow.

This fur is peculiarly constructed. Next the skin it is woolly and somewhat oily, very close-set. If the wind blows it back, or you stroke it in the wrong direction, you cannot see skin in between. The outer pelage for some distance at the end of every hair is hollow like a tiny quill. This air space, in swimming, acts like a life preserver! In spite of the immense weight of the heavy antlers, the rump of the reindeer never sinks below the water. A herd of reindeer or caribou migrating down a waterway are indeed a strange sight. They generally swim in Indian file. At the front end of every deer there rises above the water its long head with the huge branching antlers; about six feet behind this at the rear end of the animal, rises above the water the white rump and the short tail held stiffly erect.

Reindeer could almost be called amphibious, for they are almost equally at home in the water and on land. They are strong, fast swimmers and have often been observed to use a stream as a migration path where there was perfectly usable land on either side of the stream. The fawns, when they first go into the water, seem to be ill at ease and do not swim well. There is a story that in one autumn migration where thousands of reindeer were crossing a body of water, the calves crossed by the simple method of walking over the backs of the old ones as on a bridge. This is under suspicion of being a "tall story"!

We go South in the winter to escape the severe cold, and to the mountains in summer to escape the severe heat. The reindeer do this too. They migrate regularly twice a year and this migration is a very impressive sight. By the end of autumn, the stag has his full growth of antlers and the velvet has peeled off, leaving the antlers glossy and beautiful. His winter coat has come with its creamy white, long mane down the front of the throat; the rest of it a warm grayish-brown, very thick and rich. Frosts begin and the plants on which the reindeer have been feeding are killed. Finally the streams begin to freeze. When this happens the deer begin their southward migration in earnest. During the summer they have been more or less unsociable with each other. The does have stuck close to their young; the stags have wandered about singly or in twos and threes. Now large herds appear on the regular well-trodden migration routes. These routes are so established that people can watch with certainty at points on them, sure of seeing the processions of antlered animals coming in hundreds down the route. Over the herd, and trailing out on the air even above a single moving animal, there is a heavy white mist. This is the evaporation of moisture through the skin, in the form of steam. Above a resting herd, it stands like a heavy, motionless, white cloud. Accompanying them also is the characteristic click of the reindeer foot.

Usually there is a leader; sometimes two leaders. This is usually a doe who has not borne young, although it may be an old buck. They do not go on steadily toward the better feeding grounds of their more southern range. When the weather is sunny and warmish they often spend the day lazily in the sun; or they may stop for some time in a place where food is plentiful. The reindeer moss of which they are so fond, is a rather pretty, short, branching lichen, growing in large areas on the ground. Wherever they find a good field of this, they will stop and graze.

During this time the reindeer foot is becoming even

more excellently adapted for the winter conditions. This foot has the sort of hoof usually called "cloven." The Devil is also portrayed with this sort of hoof. What the Devil's hoof is like remains a topic for guesswork, but the reindeer's hoof is not a hoof, it is two hoofs; each with its own toe. The hoof is merely an enlarged toenail exactly as in the case of the four hoofs of the foot of the hippopotamus or the one hoof of the horse. The tracks typically made by the reindeer foot show the nature of its outer structure, but of course do not show that inside the two halves, widely separated in the middle, are two separate toes. These tracks in soft ground or snow look like two half moons, facing each other, with space in between, the points of the crescents sometimes nearly touching. Behind these at a distance of a few inches are two elongated, triangular marks, each a little to one side of the line from the one in front of it. These latter show most plainly in soft muck, swampy land or snow. They are the marks of the small accessory hoofs—the "cloots," which grow pointing downward behind the other hoof and just miss the ground, unless the ground is soft enough for the front hoof to sink in a bit, thus bringing the cloot to the level of the ground. The front hoof in the warmer season has a quite fat "frog," or pad, beneath it. Toward the end of the autumn this frog is gradually absorbed, leaving the naturally sharp edges of the hoofs even sharper. This gives the animal a very secure purchase on the icy slopes it must often travel. On the under side of the hoofs grow long, stiff hairs filling in the space between the hoofs and giving the foot additional hold. This formation also aids the animal in swimming, acting as a paddle and possibly keeps the foot from "balling," as a horse's hoof does when it is traveling over soft or wet snow.

On mucky or swampy land in which a differently built animal would be mired, the reindeer lopes along on its spreading hoofs and its slender strong legs in swift, even strides, the head outstretched as it runs. In very soft mire or snow, the whole leg—hoof, cloot, up to the hock—is used strongly to propel the animal forward.

Traces of its passing are sometimes left in the form of a fluid, secreted by a gland in the foot. Possibly the smell of this fluid is a warning signal, for it is most apt to be secreted when the animal is alarmed.

This is not the reindeer's only warning signal. When a herd is grazing or at rest, there is more than often a sentinel animal, and this animal, like the sentinel chamois, bugles a note at the approach of danger. Unlike the chamois, however, the watchman of the reindeer herd is usually a female, not a buck.

The reindeer's feet have another and a very puzzling peculiarity. This is the clicking noise one hears whenever a reindeer moves, and even when it is standing still. A group or a herd of reindeer on the move make a sound like dozens of castenets being quickly and lightly clicked together. The first explanation offered for this noise was that the "cloven hoof," which spreads when it touches the ground, clicks together again the minute it leaves the ground. However, the sound usually occurs when the foot is on the ground! By inducing a domesticated reindeer to take a stroll beside him, one investigator found that the click came from a region nearer the cloot than the hoof in front of it. The latest theory advanced by many observers is that this click is made on the inside not the outside of the foot and that it is caused by a tendon slipping sharply over a bone.

At any rate, it prevents the reindeer from approaching anything without warning of its arrival, for it is always with the animal and is one of the few sounds the animal makes. Occasionally a reindeer bugles to warn of danger; in running fast they make a coughing pant. The cow and calves grunt at each other, and in the breeding season the males produce a guttural coughing bark or grunt, which the Indians and Eskimos can imitate very perfectly. Reindeer also are very noisy eaters! As they chew, they gnash their cheek teeth loudly together. This sound is fairly frequently mistaken for the clicking of their feet. Other voices they do not seem to indulge in, and even in great pain they are mute.

Behind the migrating herds streams freeze and the heavy snow blots out the land. And under this snow the reindeer's food lies buried, except when a hard wind blows the snow away from a patch of ground and the lichens are exposed. The animals do not suffer nearly as much from the terrific cold as from a lack of food. Their soft dense fur with its hollow ends provides an almost perfect form of insulation. It retains the body heat, allows evaporation of moisture, and sloughs off snow and rain. Their large spreading hoofs act like snowshoes, and the sharp edges of the hoofs, together with the hairs beneath the foot, prevent the animal from slipping. These sharp-edged hoofs are also used to scrape aside snow, in order to get at moss beneath it. Many people think the brow antlers are also used for this purpose, but the opinion does not seem to be substantiated. As a matter of fact, these antlers are lost before the season of heavy snowfall and starvation begins. The stag who has spent five months of the year growing his handsome antlers, goes through all this period of growing them only to lose them at the beginning of winter. He really has very little use out of these decorations, for while they are growing they are covered with sensitive velvet and have to be guarded from anything that would be painful to the velvet. After the velvet has come off, the antlers rapidly begin to dry and in a short time they fall off. The doe, although she is the only deer doe that has any antlers at all, has much smaller ones than the stag, and carries them much longer.

The intense cold of the Arctic winter finds the reindeer in the forests where protection and nourishment can be procured. Here they spend the winter, wandering wherever they can find food.

When the snow softens and the ice begins to break, the return migration starts. The rivers are torrents of rushing snow water, filled with blocks of ice; every mountain side is a glistening cascade of water. This is a very different migration from the compact, orderly march of the autumn. And the animal is also far from the compact, orderly and well-groomed animal he was in the autumn. His winter coat has begun to come out and he looks very untidy and rough. The does, who keep their little horns nearly all winter, have not lost them and the stags have begun to grow theirs again. They do not go back in the herds they went down in. The does are in a hurry to get to the cooler regions before their young are born. They leave ahead of the stags, in small groups. The stags come along later in small herds. It is perhaps even more of a battle to get North than it was to reach the South in the cold season. Plowing through flooded swamps, swimming the bursting torrent of ice-cold mountain streams, the animals manage to reach their summer home before spring has ended. Again they can be counted on to take certain definite migration routes, in spite of the heavy obstacles they are bound to encounter. At some northern railroad crossings, for instance, the inhabitants know the approximate date of the reindeer's appearance, and watch for them with the certainty that about that time the herds will begin to cross the tracks.

The spring migration of does is widely separated from that of the stags, because of the hurry of the does and the lack of haste of the stags. In some Arctic regions the stags do not migrate at all; only the does and fawns.

So the reindeer has gone from its summer home in the less protected, cooler regions to its winter home in the more protected and comparatively warmer regions. It has changed its covering to fit the climate; grown its massive antlers and lost them; gone through its season of plenty and its season of famine; borne its young; made its return migration, and is now back in the summertime again. It has completed the cycle it will go through year after year for the twenty years or so of its life.

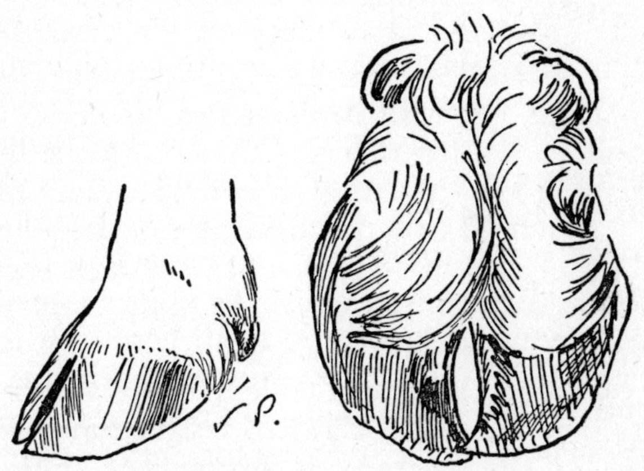

THE CLOVEN HOOF (p. 296)
(redrawn from Jacobi)

18

THE BEAR

DAY in, day out, James Capen Adams sat at his shoemaker's last, cutting and stitching leather, shaping it into shoes for the good folk of the town of Medway, Massachusetts, wishing at every stroke that he had work more to his liking.

Then, one day, when he was about twenty-one, a traveling wild-animal show came to town, and set up its cages. It was a thrilling event, more exciting than anything the villagers had ever known, for a hundred years ago such spectacles were rare. When the show left, James Adams left with it, deserting his bench, last and tools, and everything that had to do with the humdrum of making shoes. At last he had work nearer his heart's desire. As official collector of wild animals for the wandering menagerie, he was to add to it such creatures as he could capture in the woods of Maine and Vermont, following the route of the show.

His career as a showman ended abruptly, however, for a severe mauling at the claws of a tiger forced him back again to the dreary trade of shoemaker. This time he set up shop in St. Louis. Luck was not with him, and once again he put away his cobbler's tools. This was in the days of the forty-niners, when the air hummed with talk of fortunes and adventures waiting for any man of spirit in the Golden West. Adams rolled up his kit, and started for California, the "land of promise." He lived in the open. He tried his hand at mining, but without success. He turned to stock-raising, and failed at that. Next he tried trading among the pioneers and panhandlers, but he was no business man, and he soon discovered it. In a mood of discouragement with himself and his fellow men, he forswore their company and competition; took to the mountains, and became a hermit.

Armed with his gun and relying on his young brawn and muscle, Adams began anew. There in his solitary camp in the hills, the recollections of his days with the traveling animal show awoke, and he decided to take up again the work of collecting wild animals for sale.

There was one animal in the California mountains, however, which Adams made no effort to capture, a hermit like himself, the dreaded grizzly bear. He was a killer, men said, a ferocious, bloodthirsty brute that unfailingly charged in a rush attack, without provocation, seizing his victim and hugging him to death, or biting with his iron jaws and tearing his prey to shreds with the long claws of his powerful feet.

Warned that a gruesome death was the sure reward of any man who crossed the trail of the grizzly, Adams took good care to avoid him. The grizzly returned the compliment. Occasionally, Adams caught sight of a big, dark form among the trees, and once in a while took a pot shot at it, from a safe distance, but he steered clear of a face-to-face encounter.

A few years later, the same man walked through the streets of San Francisco followed by a troop of the very bears that had so terrorized him when he first took to the woods! His grizzlies padded along after him, unchained, as meek as Mary's little lamb, paying no attention to the crowds of men, boys and barking dogs that trailed after them.

By this time the shoemaker-hermit had a new name and a new fame. He was hailed everywhere as "Grizzly Adams," and as Grizzly Adams he has gone down in history—the man who best knew and best understood the dreaded bear of the Rockies.

He was by no means the first to report on the grizzly. That distinction fell to Lewis and Clark, two army men who pioneered across the continent leading an expedition authorized by President Thomas Jefferson, for the purpose of reconnoitering the country that stretched from the Mississippi to the Pacific. They were the first white men to make the overland journey north of Mexico, and they blazed the trail to the coast, opening the way for the settlement of the West. Lewis and Clark made friends with the Indians wherever possible, seeking to gather from their talk with the Red Men information that would guide them through the unknown territory that lay ahead.

In the Indian wigwams they heard stirring tales of the hunting of the *hohost,* or "white bear," a bear different from all other bears; dangerous to the last degree. No Indian dared face him single-handed. They set out in bands of six or eight, armed with bow and arrow, painted as for war, their courage stimulated by the same dances and ceremonies that preceded battle with an enemy tribe. The braves who came back unharmed and victorious were greeted with shouts of acclaim. As trophies of the hunt, they wore necklaces strung with the white bear's claws, and any Indian so decorated could be sure of admiration wherever he went.

Lewis and Clark had plenty of opportunity to verify the tales of the grizzly as a fighter. The expedition was camping on the shore of the upper Missouri, just at the mouth of the Yellowstone River, when their first personal acquaintance with the "white bear" began. One evening shortly after dusk, Captain Lewis and one of his hunters caught sight of two grizzlies coming toward them. Both men raised their guns. One of the wounded bears made off. The other came at Captain Lewis, but

GRIZZLY BEAR

his speed was slackened by the pain of his wound, and after two more shots, he dropped.

The explorers had a number of hair-raising encounters with the grizzly. They had learned from the Indians that, unlike other bears, he cannot climb, and more than once they took to the branches for safety. At other times, the escape was made by a plunge into the river, where the bear was handicapped by having to swim after his victim, so that the hunters on shore had time to shoot before the trapped man could be fatally caught. In many instances, however, the animal disappeared without the ceremony of attack, seeming as anxious to avoid a meeting as the men themselves. He fought with desperate ferocity when molested or alarmed, but otherwise he kept to his hermit-like solitude.

Why the Indians persisted in calling him the "white bear" was not clear to the expedition men, for when they showed the Red Men bearskins that were black, brown or red in poil, the Indians sorted out certain ones of them and classed them as hohost skins without hesitation. All those that had hair pointed with white or frosty tips were "white bear," they explained. So, in the language of the day, those with a blur of gray points to their hair were described as "grizzled." In later days, as the trappers and settlers came to know the grizzly better, they thought up a variety of other names for him. Because of the dense half-ring of thick fur that rose around his neck and shoulders like the clipped mane, or roach, of a horse, they named him the roach bear. If the bear appeared to have a whitish muzzle, they called him "bald face." As their terror of him decreased, they indulged in more familiar nicknames for him, such as Silvertip, Old Ephriam, and Moccasin Joe. Of them all, the name that clung was Grizzly.

Unfortunately, another word, of similar sound but of another meaning, sometimes took its place—the word grisly, and this bear was often spoken of as the Grisly, which means gruesome. Though it began with a mistake, many people thought grisly the better term of the two. The man who gave old silvertip his scientific name was of this opinion, for in choosing a Latin name to distinguish him from other bears, he decided to call him *Ursus horribilis*.

James Adams was no Latin scholar and had probably never heard of the grizzly under his frightening name *horribilis,* but when he first went to live in the hills, he had the same horror of the animal that every one else had. By the end of his first year in the woods he had erased this nightmare picture of the grizzly from his mind! He had come to regard him not as a dreadful monster, thirsting to kill, but simply as a wild creature, shy and timid, possessed of the same impulse to withdraw from a strange presence that all wild things have. No one understood this better than Adams, because he himself shared it.

One day, on a trail in the mountains of eastern Washington, Adams shot a grizzly mother who was accompanied by two cubs. He adopted the orphaned bear children, and one of them, whom he named Lady Washington, became his inseparable companion. She was about a year old when Adams took her, and had already learned enough of the joys of freedom to object to his discipline, but gradually she submitted to his control. He trained her to obedience so thoroughly that he could safely take her along as companion on his hunting trips, and she helped him by carrying the day's trophies all the way to camp on her back. Lady Washington needed no chains, though she grew to enormous strength. She shared Adams' cabin at night, and he lay between her and the fire, warm on both sides.

Eventually, the couple became a trio, for on one of his hunting trips, another cub was taken. This one was so young that a foster mother had to be found for it, and the baby grizzly was nursed by a greyhound. When Adams picked up this tiny little affair of a bear, he held it inside his shirt front to keep it warm. Though a full-grown grizzly may weigh anywhere from three hundred to a thousand pounds, he begins life as a morsel of only about half a pound. At birth the cub is no bigger than a gray squirrel, quite bare of fur and blind, and though he develops to a length of about nine feet, he is no more than that many inches long when he is born.

DANCING BEAR

The tiny cub was christened Ben Franklin, and, like his namesake, he grew to high fame. Ben was devoted to his master and showed as much affection for him as the most faithful of dogs. He was always at Adams' heels. He showed his loyalty more than once, even to the point of engaging in battle against one of his own kind. Adams was once attacked by a big grizzly and in extreme danger, when in came Ben Franklin to the rescue, fighting claw to claw, though he was badly wounded himself, to defend his master. There was no measure to the love that Adams had for his cherished grizzly.

Later on, Funny Joe, another cub, was adopted, and, finally, Lady Washington had a son, whom Adams named Fremont. Fremont was the least promising of the bear babies, less pleasant and less intelligent than the others. The giant of the group was Samson, whom Adams captured when he was nearly full-grown. Samson could never be trusted to go unchained as the others did, but he was an interesting proof that even a grown-up grizzly could be made to respond to training and kindness.

After a stay of several years in the Rockies, Adams decided to return to the East. He took passage for himself and his bears on a sailing vessel, and together they rounded Cape Horn. But Ben Franklin was not of the party. He had died. Adams grieved for him as though he had lost his best friend, as indeed he had.

On their arrival in New York, the greatest of showmen, P. T. Barnum, engaged the Grizzly Family, including Adams, to appear at his circus. There the fearless shoemaker-hunter did a great service to his shaggy friends. People flocked to see with their own eyes the bear of bears, the murderous grizzly. They came expecting to see powerful brutes, heavily chained, growling under the whip, teeth bared, claws ready to rip. Instead, they found a harmonious family of huge silvertips, playful and thoroughly tame. Then and there the grizzly's horrible reputation began to dim.

Though disillusioned as to his savage ferocity, their interest in the bear of the Rockies did not wane. Wherever the circus pitched its tents, crowds gathered around Grizzly Adams and his troop. The good old trick of sneaking in under the edge of the "Big Top" was as well known then as it is now, and boys crawled in for a stolen glimpse of the wonderful man and his wonderful animals.

Everywhere and always, the bear has inspired a special interest and amusement. Back in Biblical days, young and old alike enjoyed hearing about the thickly-furred animal that stood up comically on his hind legs and walked upright as they did. They discovered that he had a weakness for sweets even stronger than their own, and stole grapes and pilfered the fruit and nut trees as he clambered from hill to hill. It was the Syrian bear that they knew of, the same bear that now roams about the forested slopes all the way between the Caucasus Mountains and Palestine.

The youngsters of cultured Greece clamored for stories about the brown bear of the dark Peloponnesian hills. The Romans knew of his presence in the forests of their own country and of their conquered provinces. And at the same time, the savage Indian boys and girls of the Andes sat in the sun or clustered around the night fires, just as they do to-day, exchanging stories of the little black bear with the broad, short head and a yellow half-moon over each eye—the spectacled bear of South America. While in the Far East the Malays have a little bear of their own—the funny little Bruang, or sun-bear—which shuffles about in the warm woods of Malaya and Java, Sumatra and Borneo. He is small and only about a yard long, with a yellowish blotch on his throat. He walks upright more than any other bear. Because he is sweet-tempered and amusing, he is often kept as a pet, and as he prefers honey and sugary fruits to other diets, he is easy to feed.

An altogether different creature, especially in disposition, is the sloth bear of India. He too lives on sweets, supplemented with a diet of insects which he licks up with his long tongue and scoops from the ground with his flexible, funnel-like lips. But the sloth bear is regarded as by far the most dangerous animal in the Indian forests, and the natives are in terror of him. He is uncertain in his behavior and it is a toss-up whether he will quietly shamble away when startled, or dash out in a headlong rush in deadly attack. The mother sloth bear, if frightened into retreat, carries her offspring with her pickaback.

In most countries, however, the bear inspires respect rather than fear. In Japan, the hairy Ainos, who still cling to their ancient tribal mode of life, worshiping the sun, moon and sea and other phenomena of nature, hold a festive holiday every year in honor of the bear. The fisherfolk of the Amur District in Siberia find him worthy of serving as a sacrifice to one of their important deities. Our own Modoc Indians, of California, believed that their tribe was descended from bears, and therefore never killed one of the revered animals without begging forgiveness for the sacrilege to their ancestors.

As a historical figure, Bruin is immortalized in Switzerland, particularly in the city of Bern, which is said to have been founded on the site of a thrilling bear hunt. The municipal coat-of-arms shows a bear standing upright, and the statue of the national hero, Erich, represents him as guarded by four of the animals. The city maintains a huge bear pit which is always surrounded by spectators.

Russia's bear is supposed to typify the national character, and represents that country in cartoons, as the eagle does the United States.

Almost every child in Europe comes to know the form and antics of Bruin Blackfoot, the dancing bear, just as we know the organ-grinder's monkey, for wandering bear trainers make the rounds, stopping at the squares and street corners to give their performances. This custom is very popular in England, where it has survived since the Middle Ages, when bears were brought from castle to castle, to furnish entertainment at banquets. Whenever a tournament was held, tricksters and motley performers gave their shows in a nearby field, and there was always a troop of trained bears to amuse the crowd. It was the European brown bear that supplied the fun, as he does nowadays, though sometimes there may be one or two Himalayan bears, distinguished by the white half moon at the base of the throat, in the troop.

While the good-natured people laughed at the tricks of the performing bears, there were others who found pleasure in the hideous spectacle of bear-baiting. A pack of dogs were let loose on a shackled bear, and the sport of watching the fight delighted the onlookers much as bull-fighting fascinates the people of Spain. By an Act of Parliament bear-baiting was finally forbidden.

The brown bear of England as a wild creature came to an end at the beginning of the twelfth century, and elsewhere in Europe he also died out with the coming of civilization. He is still found in the Pyrenees and in the eastern Alps and thence to Russia, Syria and northern Asia. Bear-hunting in all these regions is practiced more for the sake of sport than for any useful purpose, although in northeast Asia, where it is very cold, the wilder tribes keep warm in bearskin robes and grow fat on bear meat.

Of all the members of the tribe of Bruin, the handsomest is the polar bear, whose home is on the ice floes in the Arctic regions of both hemispheres. His long, pointed head and slender neck, as well as his uniformly creamy coat of fur, set him apart from all others. The color of his fur does not vary with the coming and going of winter, but remains white all the year round. Another of his special features is the stiff hair that covers the soles of his feet, by virtue of which he can get a foothold on the slippery ice. His claws are shorter than

POLAR BEAR

those of other large bears, and well surrounded with hair, so that his tread is completely muffled. He can steal up silently to attack seals as they lie basking in the sun. His ivory-colored fur is valued by the Eskimos for its warmth, and by the rest of the world for its beauty. Priests in the Arctic spread the whitest skins they can get as rugs before their altars.

No part of the world has as many varieties of bear as North America. At first only the several varieties of black bear were known, differing slightly in skull formation and other details, but in general the same in appearance and habits. This bear is usually black; brownish at the muzzle, and with occasionally a white patch on the breast. His fur is smooth and fairly glossy. He is rather small in the head and his hind feet are much shorter than his forefeet. He is not a very big creature altogether, and five feet long is a big size for him to attain. At one time the whole continent from Labrador and Canada down to the Gulf of Mexico and all the way from the Atlantic to the Pacific, sheltered the common black bear, but he has been much reduced in numbers and in many parts completely exterminated.

The black bear is not a formidable creature to deal with except when wounded or fighting to defend its cubs, but he proved a great nuisance to the farmers of earlier days who objected to his stealing honey and robbing their orchards and vegetable patches, and even more to the occasional abduction of their fattest and roundest pigs. The hungry pirate was therefore driven and hunted farther and farther into the wild, and it is only in regions remote from settled country that he is to be found. Much as they resented his raids on their pigstyes, the young farmers and villagers did not despise his fat as an aid to beauty. Of a Saturday night, dressing in their best to drive to town or to take their best girls riding, they would slick down their hair with a touch of bear's grease, feeling that the elegance of the pomade made up for its smell. The old folk dipped their fingers into the same pot of bear's grease, to rub it into their aching joints as a cure for rheumatism.

In the days when people rode in carriages, every one had a thick bearskin lap-robe for the winter. They are still to be seen in the few hansom cabs which stand, as in the old days, at one of the entrances to Central Park, in New York City.

The handsomest of our black bears often come to a proud end in the form of the high bearskin caps, or busbies, worn by the aristocratic Grenadier Guards, of the British Army. One bearskin makes two caps. These are held on by will power and a black leather chin strap, and are considered a most swagger adjunct to the blue trousers and red tunic of the elegant guardsman.

But the black bear is not always black. This is particularly true in the West. He often appears there with a yellowish coat under the name of cinnamon bear. Like the blue-gray glacier bear of Alaska, the cinnamon bear was formerly supposed to be a distinct species. Now both of these are usually included with the black bear, especially since cubs of the same mother have been found to vary in color.

The two bears of America which are distinctly different from the Black, are the Grizzly and the Alaskan Brown Bear. The first explorers of the West ranked the Grizzly as the biggest bear in the world. Bear-hunters came back with swashbuckling stories of the giants they had brought down, huge and massive, so powerful that it took a dozen bullets to lay the animal low. With every telling, the grizzly grew in size.

Then came the discovery of the Alaskan Brown Bear, and the grizzly's nose was put out of joint as a heavyweight. Although Alaska had been a United States possession since 1867, when it was purchased from Russia, it was not until 1898 that her brown bear came into the limelight of natural history. The peninsula, with its high, rock-bound coast pointing far out into the Pacific, has an area larger than all of New England plus the Great Lakes and the whole of the Atlantic Coast States, except Georgia and Florida. There was plenty of talk of salmon and seals and of timber and mines, but the huge brown bears of Alaska escaped general notice all this time because they lived away from the centers of these industries. As the interior gradually opened to travel and sport, the presence of the brown bear began to be widely known. Comparisons at once began between him and the grizzly on the point of size, and the decision was unanimously in favor of the Alaskan bear as the larger. Not only is he the largest of bears, but the largest carnivorous animal in the world. He stands as high as fifty-two inches at the shoulder (the she-bears are smaller); his haunches and hind legs are massively built, and he weighs as high as fifteen or sixteen hundred pounds. His rear part is so enormous that the outline of his hind legs is almost shapeless.

There are grizzlies in this same territory, so that a hunter is not always sure which of the two he is following. But an experienced hunter can usually tell one from the other by his gait. The grizzly's walk is much more easy and leisurely. He moves his great body with less evident effort and with more grace than the lumbering brown bear. When in a hurry, both of them can move with great speed.

The hind paws of all bears are strikingly different from the forepaws, and greatly affect their manner of walking. They are nearly as plantigrade as our own feet. The sharp indentation behind the ball of the foot and a pointed heel are the special marks of the grizzly, quite different from those of the black bear, for instance, whose feet are narrower across the sole and more rounded at the heel.

Bears have a habit, very practical in deep snow, of walking in each other's footprints, just as the children who have forgotten their overshoes carefully walk in the footprints of some one who has his on. This makes a very definite and lasting trail. But the grizzly bear's trail, even where there is no snow, is marked in another way—by the patches on trees where, standing on his hind legs, he has, for some unknown reason, scratched off the bark.

The grizzly has longer foreclaws than any other kind of bear and uses them with much more skill. The front claws—about twice the length of the hind ones—are from three to six inches long, and not only terribly powerful, but extremely agile and sure in their clutch. The grizzly uses them to turn up stones and logs, where he suspects he may find food, for grubbing after insects as well as for fighting weapons. The claws of the grizzly are sometimes white; sometimes black, and sometimes striped. They are usually described as "nearly straight" because they do not curve as abruptly as the claws of other bears, but as a matter of fact they are often quite arched, although individuals vary in this respect as they do in so many others. In the early spring, after months of disuse, they are at their longest. After several months of scraping the ground, digging and scratching, they become worn and blunt. A bear's claws are fixed, that is, quite without the retractile power of the claws of the cat tribe, so that they are exposed to wear all the time. But they are strong even when they are blunt, and can do the work of an iron rake. They can also do the work of a delicate pair of tweezers, and a grizzly can pick up tiny morsels by pressing two claws together in a very fine clutch. He never wastes his claws, however, and if he can use a single one in scooping out a tidbit, he makes that one do.

The shape of the bear's feet, which enables him to

set them flat on the ground after the manner of man rather than of quadrupeds, enables the bear to stand erect as we do. Some bears do this much more habitually than others. When the grizzly is alarmed or curious, he rears up on his hind legs with his body straight, so that he can get a clear view. The Alaskan brown bear rarely takes this posture, but quite frequently sits on his haunches like a dog.

No other animal, not even man, has as varied taste in food as the grizzly. There is hardly a thing, vegetable or animal, in his world that does not please his palate. Whereas other animals travel far and wide in search of their chosen food, even to the extent of prolonged migrations from season to season, the grizzly stays at home, and when one form of food gives out, he changes his menu, adapting his taste to whatever he can find. He will travel some twenty miles or so to satisfy his appetite, but ranges no farther.

Other bears manage to confine themselves to a vegetable and insect diet, or feed on honey. In hot countries, they are sure of food the year round. In Europe and the eastern United States, when the summer fruits are exhausted, and the winter frost denies them their usual fare, they fatten up before the cold weather and sleep through the months of food shortage.

The polar bear, in his home on the ice floes, lives on the food that the Arctic seas supply. He is a good fisherman and a powerful swimmer and follows the seals and walruses. He is heavily padded with fat under his white coat, and though a heavy creature, in the water he is as light as cork. He changes his beat with the floating ice packs, always following where his prey goes. In the summer when the ice floes break up, the ice bear often makes for the shore, and there he crops the grass as greedily as any herbivorous animal.

But the grizzly keeps to his own territory the whole year round. He is usually out on the food hunt late in the day, although in remote regions, where he feels himself secure, he is sometimes on the prowl while the sun is high. He eats anything and everything. Herbs, fruits, flowers, leaves, bark, roots, grubs and worms, every sort of insect, even ants and wasps, their larvæ and their nests; he robs the bees of their honey, unafraid of their sting, for his thick fur protects him. Together with this miscellany, he feeds on such delicacies as strawberries, clover, mushrooms and anything that bush or tree affords. He is no respecter of private property, and many an unlucky squirrel has had his store of nuts devoured by a passing grizzly. He will even eat the squirrel itself, if he can lay his paws on it. A powerful grizzly when hungry will devour gophers, marmots, an occasional porcupine—in fact any animal of any size from mice to moose.

Old Silvertip has an extraordinarily keen sense of smell, much finer than that of a dog, and it leads him to the discovery of hidden dainties cached in crevices and under logs and stones. It leads him also to the stores of food in camp, where hunters have their hideaways, and bears will invade camps where they scent a supply of dried apples, sugar and especially bacon. One such lure brought a grizzly into camp one night where a man lay asleep. The grizzly attacked him ferociously. This was cited as an instance of the grizzly's tendency to attack unprovoked, for certainly a sleeping man could not be said to have aroused the bear to ferocity. Later, however, it was discovered that the man had made a pillow of his bacon pack in order to keep it safe!

In the days when the buffalo roamed the plains, grizzlies were at home on the edge of the prairies and often came out in the open to make sure of a buffalo steak. With the settlement of the plains, of course, they disappeared. Since then, they have been guilty of sheep-stealing, and though it is not their common practice, an occasional silvertip makes a raid on a flock and dines off mutton when he can. This is an exceptional feast for him however. He will feed on meat that he has not killed, as well, and whether it is fresh or not seems to make no difference to him. If he is spared the trouble of killing it, so much the better. It is his characteristic to save what he can for future use, and often he carries off the remainders of his feast and buries them under a pile of scraped-up earth.

The grizzly's fondness for meat has resulted in his undoing, for even when he retreated from the traveled routes, he was occasionally tempted down toward a sheep ranch, and because of trespass on the part of a single grizzly now and then, the whole tribe became branded with the name of sheep-killer and was forfeit to the hunter and the ranchman. Men out for sport used this as a pretext for a wholesale slaughter of the grizzly. Men who wanted a good, warm bearskin or a day's sport defended their hunting by saying that they had removed one more menace to man and cattle, and they still urge the extermination of old silvertip on the same grounds.

In the old days, the pioneers of California who went out into the hills caught sight of as many as thirty and forty bears in one day's tramp; whereas in 1932 there were only eight hundred grizzlies left throughout their range.

The best feeding season of all for the bears of the Rockies is the time when the streams are filled with fish, especially with salmon. In the spring, the salmon come up the streams from the sea, to find clear, quiet water in which to spawn. This urge, combined with their powerful muscles, leads them crowding up the Alaskan streams, leaping over falls and braving the rapids in their rush to get up to the spawning grounds. Often they land there very well bruised, and sometimes they are killed. After spawning they die, and the young salmon go down the streams to the sea, thus beginning the life cycle over again. During this time the shores of the streams are filled with dead fish and the sea gulls fly miles inland to feed upon them.

The bears do not wait for the salmon as a ready-made catch. They are fishermen of the highest order. A little before the salmon rush, they come down from the high altitudes of the rough timber hills and take up residence on the river banks. When the salmon begin to appear, the grizzly takes his place, squatting on the shore, or sometimes settled on a log out in the stream. As the salmon speed by, out sweeps his mighty paw, and with a scooping slap, he throws the twisting fish in a great shower of spray on the shore. Sometimes the salmon drops, after a neat parabola, onto the ground as much as fifteen feet away from the bank. Grizzly feeds his first hunger with the first of his catch, but goes right on fishing, like any other enthusiast of the art, and often piles up a dozen or two for future reference. It is his custom to cover his store of food, at least partially, and he usually scratches up enough earth to conceal his hoard, burying his leftovers with the same instinct that prompts a good housewife to line her cellar shelves in the canning season. After the fishing season and during it, the bear smells of fish from head to tail, and even after the long winter, he reeks of it.

By the end of October, after a rich feeding on salmon

and the sundries of the summer diet, the grizzly is thick with fat. With the coming of the cold weather, the small creatures among the rocks and roots of the hillsides vanish and the fruits and berries are gone. Then, as though with the disappearance of his food supply his last interest in life were gone, the grizzly refuses to bother with all activity. As long as there is still something to eat, he remains on the scene, and in regions where there is enough to eat, he sometimes stays abroad throughout the winter. As a rule, however, he solves the hunger problem by going to sleep and forgetting it. When the frost of fall begins to nip, he prepares a den, either selecting a convenient cave or digging one out in a hillside or under a clump of roots. The spot for the dugout, wherever it may be, is always well drained. For a month or two before he retires to his den, his appetite is slack. He has eaten so much that he can hold no more, and though he does not move into his winter quarters immediately, he does very little feeding.

In the Far North, many animals carry on a regular migration when the cold weather comes, and their food supply begins to be cut off. The bear is not driven to this necessity, as he is practically omnivorous and can get food when other animals cannot. The male Polar Bear simply extends his range, going out farther seaward to the edge of the ice, and there carries on his fishing, sealing and walrus-hunting. The refuge of the bear in face of lack of food is his hibernation, or winter sleep. As a matter of fact, it is much more a stupor than a sleep. Having gorged to the bulging point, he finally withdraws to the den. The female ice bear simply removes herself from the ice pack and makes herself a cozy nest in a snow drift. There she bears her young and nurses them until she and they emerge in the pale spring sun.

In the North there is still heavy snow in March, and the hibernating bears may not emerge until April; the females and cubs not until May.

Nature adjusts the bear's inner mechanism marvelously for the winter retreat. All the natural processes become sluggish to the point of stopping. Respiration and circulation go on, but sluggishly, and elimination, except by the lungs, appears to cease entirely. The animal is living on its own stored-up fat. When it leaves the den it is no longer the bulgy creature it was when it disappeared from the outer world. It looks hungry.

But instinct seems to direct its activities at this time, for the bear does not plunge out into an eating orgy at once. Like a person who has starved, it must be very careful how it eats for a while. First, it makes for a drink of water, and then feeds sparingly on very light food, either grass or young shoots. Again instinct seems to guide it in the right direction, for if the bear is near the coast, it is very apt to take a dose of kelp, which is very cleansing to its system. After a few days of cautious diet, the heavy feeding of the summer begins with abandon.

The winter den is the first home of the cubs, and the mother bear, having them to care for, never sinks into a full stupor. She times her exit from the den by their strength, and waits until they are able to toddle after her before introducing them to the outside world.

At about three months of age, when they first come out, they are the size of a cat, and weigh from ten to fifteen pounds. At the age of half a year, they have doubled in weight, and when they are two years old, just before preparing for their second season indoors, they will have grown to a couple of hundred pounds!

Baby cubs are nourished by their mother's milk for

some months after the family emerges from the den, and while the mother is busy digging up roots and crunching weeds and grasses, the cubs are quite indifferent to the business of eating, even when they are eight months old. The mother gradually accustoms them to ordinary bear food, but whether they have her example to follow or not, in due time they quite naturally take to all the things that grown-up bears love. Even bears taken when they were tiny, and fed by masters on milk and soft foods, as they grew older knew just how and where to forage for themselves. Hornaday had a black bear that had been so tended, and yet when the time came and he was loosed from his chain, he knew how to do his own grubbing, even to the extent of digging down over a foot deep to get at a root that had not begun to show above ground.

When cubs are taken from their mothers as babies, they cry for them for a good fortnight. When they are left with the mothers, they remain more or less in the crybaby stage for a good two years. By that time, both mother and cubs are ready for the separation, and the youngsters go off to establish their independent existence, usually in pairs for the first season or two.

As far as she can, the mother grizzly keeps out of the way of male bears while her babies are small, avoiding all danger of their being hurt or even eaten by them. In fact, she keeps out of the way of male bears all the time except in the mating season of the summer months, if she is wise. For the male bears do not seem to care for their lady relations, and if they encounter them, they often give them a sideswipe in passing. The family unit is not a bear trait. Nor do they go about in friendly groups, or unite for the hunt.

Polar bear adults play as well as polar bear cubs.

These big white creatures slide belly-whopper on the ice of their home haunts, or lie on their backs in the water putting hind and forefeet together and rolling over and over like barrels. Mother and cubs play together in a very jolly way. Grizzlies are not playful, but the mother behaves with her cubs very much as mothers do with children in all walks of life, even to the point of teaching them the ways proper for bears—by the spanking method. The cubs roll about in the sun, and have wrestling matches, and the mother bear permits their sprawling about as much as they please, but insists on a sort of obedience. This obedience is very necessary. Cubs are not always as quickly aware of danger as their mothers, and in one case, when, through lack of spanking, an infant grizzly had not learned to come when he was called, he took to a tree at the approach of a couple of hunters, while his mother and her more obedient offspring scuttled away to safety. The hunters caught the rebel, who thought he knew more than his parent; stuffed him into a sack and rode off to their camp, seven miles away. There they chained the unfortunate cub and put him in a cage. That night the mother grizzly padded into camp; smashed the cage; wrenched the chain loose, and carried her disobedient son away to safety! How she ever found her way over such a distance no one knows, but the instinct of mother protection must have sharpened her wits abnormally.

The grizzly's common sense grows with his bulk, and long before he is at his biggest he is a very wise and canny creature. He learns all the ways and means of living, food-hunting, den-building, and whatever else his instinct advises, but from day to day his caution increases. Grizzlies may grow up to be fighters or cowards. No two are exactly alike in color, claws, size or courage, though for the most part they are brave fighters. One of the grizzly's strongest weapons, perhaps his best, is his trick of evading pursuit. When he is followed, he does not merely make for the quickest escape; he has more deceitful tactics. From time to time, on the run, he makes a detour. He is always aware of what is behind him, but the pursuer loses the trail.

There is a constant struggle in the bear between caution and curiosity. Almost always caution wins. The bear is more than a match for most men, and it often happens that a bear hunt turns into a game of hide and seek in which the man loses both the game and his temper, but not his respect for Moccasin Joe.

As time goes on, this animal that was once only known as a killer, is becoming the representative of "courage, tenacity, and power."

19

THE OKAPI

OUR grandfathers would have thumbed the dictionary in vain if they had looked for information about the okapi, for in their day, as far as the world of books and knowledge was concerned, there was no such animal. The story of the okapi should rightly begin with "Once upon a time." It is really that kind of story. It is strangely like one of the old-fashioned fairy-tales, for it centers around a band of little people who lived in the heart of a dim, far-away forest, where green shadows crowded out the light, and where men of the outside world never ventured. Once, when a group of the little people strayed beyond the walls of their forest, an unscrupulous man caught and bound them, and threatened them with dire harm. Fortunately, another man came to their rescue, and led them back in safety to their native woods. . . . And so the okapi was discovered!

Once upon a time then (since it is that kind of story) there was a young English painter named Harry Johnston. He was talented and ambitious, and when he was eighteen years old, he left London and sailed across the English Channel to continue his studies in other lands. He loved not only painting, but architecture and languages as well, and the art and speech of other peoples interested him greatly. He traveled in Europe, and even dipped down to the shores of Africa. Wherever he went, he looked with keen eyes and an eager brain.

On his return to London, Harry Johnston wrote an account of his travels, telling particularly what he had seen in Africa. His writings won instant attention. In those days of the latter part of the nineteenth century, little was known of the land of the blacks. It was still mysterious territory. A few bold explorers, like Livingstone and Stanley, had proved that its mighty rivers were navigable and that the boundless wealth lying far within the interior could be reached. All Europe had her eyes on Africa. England, especially, was on the hunt for information that would help in the development of British colonies on that great continent. She had vast plans afoot for the further expansion of her empire, and built high hopes on Africa as a rich source of trade for the future. Harry Johnston's reports on the natives, their tribal customs and their languages caused a stir in London. His writings made his fortune. They brought him fame and a brilliant career.

In time, he became one of England's ablest colonial administrators. He was sent among the savage tribes of Africa to pacify them and to win their friendship. He used subtle tact in treating with them, and between kindness and firmness he succeeded in convincing them that England was their friend. One after another, the tribes came over to his side. Through their affection and loyalty to this one man, they bound themselves to his country, promising her allegiance and good faith and peace. One of the chiefs, Ja-ja by name, gave Johnston a hornet's nest of trouble, but he was overcome in the struggle, and there was an end to his opposition.

All the time that Johnston was so busy laying the foundations of the British Empire in untamed Africa, he found occasion for many other services to his country. His duties took him all over the land. He traveled over the plains, through jungles, and in and out of swamps. He crossed mighty mountains, he paddled up and down the sweeping rivers and across the spreading lakes of the unmapped regions. Wherever he went, he exercised his keen eyes and his eager brain. He studied the savage tribes, while he subdued them. He watched the wild game on the plains. He observed the flowers, and the trees and the countless shrubs and grasses. He collected birds and insects. Nothing was too great or too small for his notice.

Johnston made detailed reports and sketches of all that he saw. His training as a painter served marvelously to help him record in exact line and vivid color all the wonders and mysteries of Africa. He worked unceasingly. He suffered hardship and anxiety. He was laid low with fever. Often his life was in danger. But his enthusiasm and his genius stood by him, and he never wavered in his task.

In recognition of his achievements, Queen Victoria honored him with knighthood, and, as Sir Harry Hamilton Johnston, he returned to the tropical jungles to carry on his labors in behalf of the black men of Africa and the white men of the world.

One day, when Sir Harry was stationed at Entebbe, (on the northwest shore of Lake Victoria, in Central Africa), a native runner arrived, bearing a letter wrapped in leaves, fixed in the fork of a stick. The letter brought news of a shocking crime. A band of pygmies, of the Bambute tribe, had been kidnaped, and were being spirited out of the country. A European traveler had seized these pygmies on the edge of their home in the Semliki forest, in the Congo, and was carrying them off to exhibit them at the World's Fair, in Paris. The Belgian Government, which ruled over the Congo Free State, had ordered the kidnaper to release his prisoners, but he had insolently ignored the order. In defiance, he had secretly smuggled his victims across the Belgian border and concealed himself and them within British territory.

As soon as Sir Harry heard of the crime, he started to the rescue. He tracked the kidnaper and turned him over to the Belgians. He gave the pygmies a home in his compound, and nursed them back to health and strength. He could not speak their language, but he could understand their distress.

Every day he visited his pygmy guests, trying with kindly patience to discover some meaning in their strange speech. Sir Harry knew the Bantu tongue, and he hoped he might find a few words in it that were similar to the simple language of the pygmies. At last, by the use of signs and gestures and by drawing pictures, he succeeded in talking a little with them. He learned many of their secrets, and gathered a harvest of information to send back to England about these little-known people and their territory.

Sir Harry was determined to lead the pygmies back to their home in the forest. He gave orders for his *safari* to prepare, and the expedition was soon ready to set out. Food and bedding and camp equipment were all packed, and an army of porters enlisted to carry the supplies, for in those days the whole journey had to be made on foot. So Sir Harry and his *safari* and his pygmy charges began the trek toward the Semliki forest.

On the journey he peppered his wards with questions. He wanted to know how they lived, what they ate, and their methods of hunting. What he wanted most of all to learn was the truth about an animal that lived in their forest. He knew that the chimpanzee was at home in its trees, that the pygmy elephant and the dwarf buffalo lumbered through its tangled growth. But he was curious about another four-footed inhabitant of its hidden thickets. He had heard vague rumors of its existence, but had never met any man who had actually seen it.

Yes, oh, yes, said the pygmies. There was such an animal. They called it the *o'api.* It was a kind of donkey, they said, with stripes painted on it.

Sir Harry was disappointed. He thought they had not understood his questions. How could a donkey, or any kind of horse, be living in so thick and tangled a forest, where there was no open pasture? Still, it might be that some unknown variety of zebra had somehow managed to survive there. Anyway, he wanted to know more.

The journey was a hard one. As they approached the hot, steaming Semliki forest, the *safari* weakened under the strain. The choking breath of the jungle, the moisture dripping from the burdened trees, the stifling smell of rotting vegetation, all exhausted and discouraged the travelers. Then the doom of the swamp land fell on the caravan. Disease attacked them. One by one, the porters dropped under the hand of the dread black-water fever. The survivors staggered on.

Night drops suddenly in the tropics. Early one evening in the shadowy sunset, the straggling band arrived at a rest-house on the shore of a stream. Sir Harry stumbled into the shelter. His strength was gone. Bed and a blanket were all he could think of. Within, a group of black natives, naked except for a narrow cloth belted about their loins, stood shivering around the smoldering fire. They stared at the tall Englishman, and gave him a greeting. He was too weary for speech, but he raised his hand in salute. As he tottered to his cot, his eye suddenly caught a glint of white on the belt of one of the natives. Sir Harry forgot his fatigue. He forgot the fever throbbing in his veins. In a flash his mind awoke to action. Here was a clue!

He went over to the black man's side and examined the belt. Hanging from it were two little bands of hide, striped horizontally in brown and white, ending in a fringe of creamy leather. The pattern of the stripes was different from anything Sir Harry had ever seen before!

He questioned the native and learned that these "bandoliers" were a favorite ornament among the people of that region. They were hard to get, for they came from an animal that lived in the innermost depths of the forest. It was "a kind of donkey, with painted legs," said the native. The pygmies nodded their heads, in a gesture that said, "I told you so," proud of having their own description repeated.

Sir Harry bargained for the bandoliers, and stowed them carefully away in his kit. He asked the man from whom he bought them to get him a skin of one of the animals. The pygmies called it the *o'api* but the others, whose speech was clearer, called it the *okapi*.

After several days of rest, the *safari* took up its march toward the heart of the forest. When they came to the Belgian headquarters of the district, they were greeted by the white officer in charge, Lieutenant Meura. He received Sir Harry gladly, for life in an outpost is lonely, and travelers are a welcome diversion. He knew something of the okapi, and supplied Sir Harry with all the information he had. Meura had often seen the skin of the okapi, for the pygmies sometimes trap the animal with rattan nooses, hidden in a bed of leaves, for the sake of its meat. But he had never seen one alive. He, too, thought it was a kind of horse.

YOUNG OKAPI

Day after day they went out, escorted by the knowing pygmies, to prowl about in the shadowy forest in search of the "painted donkey." Not one did they see.

The little pygmies, whose eyes were accustomed to the green gloom of their jungle home, pointed to a few hoof-prints on the damp forest floor, and declared they were the marks of the okapi. Sir Harry was not satisfied. The prints they showed him were made by a cloven hoof. If the okapi were a kind of horse, his hoofs would be uncloven. No, these must be the tracks of some other animal. He felt defeated. His efforts to find the okapi were in vain.

Sir Harry was obliged to give up the search. His men were weak and ill. Many had died. He himself shook with fever. Lieutenant Meura provided him with a fresh *safari,* giving him some of his own blacks from the post, and Sir Harry turned back toward Entebbe, where he could rest and recover his strength. Meura promised faithfully to send him a complete okapi skin, as soon as one could be procured. But he died soon afterward, of the terrible black-water fever.

Meanwhile, a tiny package was sent off to London, addressed to a scientist named Sclater. It contained nothing but two little bands of hide, striped across in brown and white, and a letter from Sir Harry Johnston, telling how he had bought them from a black man at the rest-house in the Congo. These two little bands of hide became famous in natural history.

All this happened only thirty-two years ago. Men of science knew that strange specimens were constantly being discovered among fishes, birds and insects, as they are even to-day in the smaller forms of animal life. But they were certain they knew all the mammals of the earth. They never dreamed that any large animal could exist, unknown and unsuspected, as late as the twentieth century.

Dr. Sclater was astounded. He turned the two bandoliers over and over in his hands. He placed them under his microscope. He had not seen their like before. They were a puzzle to him. At the next meeting of the London Zoölogical Society, which was attended by the leading scientists of the world, he exhibited the mysterious bandoliers. They were passed from hand to hand, and scrutinized minutely. No one could tell to what animal they belonged. The scientists were baffled. They could only guess.

They finally came to the conclusion that the unknown animal was some sort of striped horse—a forest ancestor of the zebra of the plains, perhaps, still surviving in the wooded area on the Congo. They named the new discovery *Equus johnstoni,* using the Latin word for "horse" to indicate its genus, and distinguishing it by the name of its discoverer. Under this title, the stranger was welcomed on the list of the mammals of the earth.

Nearly a year passed. Then a second package came from Sir Harry. It was a large one this time, and contained the complete skin of an okapi. After the death of Lieutenant Meura, his second-in-command, a Swedish officer named Eriksson, had sent it to Sir Harry, in fulfillment of his chief's promise.

This skin stirred up a whirl of excitement among the London scientists. It proved they had made a mistake. To their consternation, they saw that "Johnston's horse" was not a horse at all!

It was not the skin itself that showed them their error, but the skull attached to it. Scientists do not make their classifications only according to the outward appearance of the animal. They base their final decisions on the structure of the skeleton and skull. When the okapi skull was examined, it was clear that the animal had no right to membership in the horse family. Instead, it was related to the giraffe!

But why, if it belonged to the giraffe family, was it without horns? This was an extremely difficult problem. But, horns or no horns, the evidence of the skull formation was proof positive that the okapi was closely allied to the giraffe. The scientists decided that it was simply a "hornless" variety.

Now, what was to be done about its name? They could not go on calling a member of the giraffe family by the name of the horse family. Announcements were sent out to all the zoölogical societies and newspapers of the world, declaring that the newly discovered animal from the Ituri forest had to be renamed. This time, the language of the pygmies supplied the animal's name. The scientists gave the word a Latin form, and the animal was rechristened *Okapia johnstoni.* Under this label it will be known forever. So the pygmies and Sir Harry, and incidentally the animal, were immortalized together.

Having the whole skin, the scientists now knew very much more about the okapi than they could learn from the two little bandoliers. In the first place, they saw that it was not really a striped animal, like the zebra. Instead, its hide was of a solid color, a purplish brown, about the shade of chocolate candy. The neck, forehead and muzzle were brown, but the long cheeks were white. Only the upper legs were striped. It was evident that the bandoliers had been cut from the hide of the hind legs, which were irregularly streaked with lines of white, tinged at the edges with a fuzz of the chocolate color, as though the dark had "run." The forelegs were also striped across with white, but in fewer and broader markings.

But what odd legs the okapi had! Not only were the forelegs longer than the hind legs (resembling the giraffe in this respect), but their striping was altogether different from that on the hind legs. From the knee down, the shanks were creamy white. Around the knees ran a circle of brown. Down the white shanks ran a narrow, straight line of brown. At the base, the forelegs decided to match the hind legs again, both having a girdle of brown around the fetlocks. All in all, the marking on the forelegs looked like the outlines of a pair of shackles, such as convicts wear, and the scientists in their reports described it as a "shackle pattern."

The hoofs they could not describe at all. In fact, they knew nothing about them. Were they cloven, like those of the giraffe and the antelope, or were they without a split, like those of the horse? No one could say. Not even Sir Harry Johnston could answer this question, for he had never seen them. Unfortunately, on the long trek from the forest to Entebbe the skin had dried and the hoofs had fallen off. When Sir Harry unpacked the skin and found the hoofs gone, he thought perhaps the rats had eaten them, but he learned that frequently, when a skin became completely dry, the horny sheath forming the hoof would drop off, leaving nothing but the bony core. The okapi's hoofs remained a riddle.

Far off in Africa, in the meantime, Sir Harry continued his government work and his research. He was not a scientist by profession, but he was fascinated by the problems of science. Naturally, his favorite problem at this time was the animal that bore his name. He tried, again and again, to get another specimen, and at last he succeeded. Again, he sent a package to London, and this time all doubts were solved.

The hoofs were blue-black and cloven, like the giraffe's. The skull and teeth resembled those of the giraffe. And in this specimen, horns were present. They, too, proved the okapi's right to membership in the giraffe family. They were not "true horns," but bone, covered with skin, as the giraffe's horns are.

The okapi was the animal hero of the day. Newspapers published the story of the arrival of the second skin in London, and reported the scientists' findings on its examination. It was mounted and exhibited. Here was an animal that had been in existence for millions and millions of years, and yet, until the year 1901 of our own era, not a single white man was ever known to have seen it.

The okapi lives only in the dim interior of the equatorial forest, on the western side of the Congo River. Until recently, no human foot, except that of the native pygmy, ever ventured into its hidden home. The tangled forest was filled with mystery—and danger. Though it was a hunting-ground for ivory, none of the stalwart Africans dared pursue the elephant into its depths. They were afraid of its pathless gloom. They had a superstitious dread of its dripping darkness. The big blacks were terrified of the tiny pygmies, whose poisoned arrows came flying at them from behind the forbidding wall of trees.

But the white man had no terror of the forest. Its mystery lured him. Fortified by his desire for knowledge, he was unafraid. The guiding genius of the scientist is his curiosity, which gnaws at mysteries until their innermost core is uncovered. As soon as the existence of the okapi was revealed, science felt a challenge to pursue the animal into the far recesses of his forest home. The skin of an okapi, studied in London, was interesting, but it was not enough. The white man wanted to see him alive!

The only way to reach the okapi was to make friends with the little human beings who were his neighbors. The pygmies have the secrets of woodcraft in their blood. They are small, their tread is light, they find their way through the pathless maze with stealthy skill. All forest dwellers, both animal and human, move noiselessly.

They know how to hush the whispering leaves and the rustling twigs. They have ears attuned to catch the faintest sounds, eyes sharpened to detect the smallest signs. They hear the silence; they see the invisible. The white man could never find his way to the okapi alone. Only the native of the woods can guide him in the misty world of trees.

In time, the white man learned to imitate the sly silence of his guide. Often he had to crawl on his hands and knees, following his little leader over the marshy ground and the lush vegetation. Their way took them into soggy swamps, through knotted grasses, over twisted roots, through a labyrinth of thicket and trees. Sometimes they spent weeks in the search. They never caught a glimpse of the okapi. He simply was not to be seen!

This does not mean that there were no okapi in the forest. But they were concealed, as if by magic. You would suppose that an animal as big as a mule, with white-streaked haunches, white shanks, and a white face would stand out sharply against the massed background of the dark trees. You would think he would be the one vivid object in such a scene. But, actually, his coloring conceals him. His chocolate coat blends into the indistinct blur of the low branches. The streaks of white on his legs are lost among the lines of the jutting ferns and twigs. The white of his face looks like a patch of lighter green in the luxuriant dark foliage. The friendly forest throws a cloak of invisibility around her creatures, and men who come to pry must train their eyes to her camouflage before they can see what it hides.

The few who came to hunt out the okapi learned not to expect a full and clear view of the animal. Instead, they studied the forest itself for betraying signs of its whereabouts. Stormy rains are frequent in this region, but on quiet days hardly a breath of wind disturbs the hanging leaves. The giant trees thrust their tall columns upward toward the sun. A shredded curtain of rope-like lianas drops from their lofty branches, with here and there a broken shaft of sunlight stealing through. The air is hot and still. Not a leaf quivers. Once in a great while, a lucky watcher might see a suspicious swaying of the lower branches. Now is the time for scrutiny. There is a chance that an invisible okapi stands there, feeding.

Men who have spent months in the Semliki forest say they have never once set eyes on a wild, roaming okapi. Some one may, of course, have come within shooting range of the shy creature, but, if so, it must have been by a chance as rare as a miracle. Collectors have manœuvred and schemed to spy upon the okapi in his home with no success. They have brought home perfect skins, but every one was obtained by native forest hunters, who caught the animals in their hidden traps.

Forest animals live in a constant state of watchfulness. Every second of their existence is keyed with instinctive fear. They are high-strung and acutely quick to catch alarm. The okapi seems to excel them all in his alertness, for it is said that he can catch the rustle of leaves a mile away.

Some years ago, the wife of one of the Belgian officials of the Congo sent a young okapi to the Antwerp zoo. It had been brought to her by a native who had snared it, and she had carefully reared it until it was strong enough to be shipped abroad. It did not long survive captivity, but since then a pair of okapi have been taken alive and safely established there. They are fed on bananas and vegetables, and apparently are able to adjust themselves to the new climate and food.

In spite of his relationship to the giraffe, the okapi does not resemble him outwardly. The okapi is only about five feet high at the shoulders, and his neck is not conspicuously long. His back has something of the sloping line of the giraffe's, but the angle is slight. The pair in the Antwerp zoo are of enormous interest, as far as their appearance and behavior are concerned, but for knowledge of the okapi's habits in the wild, the forest itself must be scanned.

The pygmies have a saying, "Every okapi pair lives in a village of their own." This springs from the evidence of their hoofprints on the forest floor. They usually wander about in twos, sometimes separating for solitary rambles. Over a stretch of miles, the tracks show that only two animals inhabit that region. They feed during the late evening and night, standing under the lower branches and stretching their pointed heads upward to reach the leaves. They are browsers, like the giraffes, and use their long, prehensile tongues to grip the leaves, cutting them with their teeth. During the day they rest. The morning hoofprints show that they make for the lower and more open clearing after sunrise, often following the shores of a brook. At sundown, the tracks point upward again, toward the slopes. You might suppose that it would be easy to follow the clue of these tracks and somehow win a sight of the evasive okapi, but he defies detection, disappearing far in advance of the approach of even the wariest pursuer.

There is something touching about the shy loneliness of this white-faced hermit of the warm woods. There is something pathetic about his random roaming and his startled retreat. And when you look at his long, pointed face, with the white cheeks framed between the cloudy muzzle and brows of purplish brown, topped with the two little peaked and pointed horns, you have a feeling that you are looking at a sad creature, on whom the singing silence of the forest and its heavy curtains of green have set a mark of despair. His eyes seem to have a hopeless, sorrowful look, like those of some unhappy spirit doomed to an eternal drifting search for something hidden.

Of course, this is only imagination. Just as easily, you might fancy the okapi as serene and content, not a driven ghost in a confusing riot of trees, but a happy spirit in an enchanted wood.

Whichever you like to think, there he lived in the distant forest of the Congo, locked away from the knowledge of man, until the eyes of Sir Harry Johnston happened to catch sight of two little striped bandoliers hanging from the belt of a black man. The okapi is our newest acquaintance in the world of wild animals, and, as with all new acquaintances, we find much to wonder at, and more to learn, concerning the secrets of his ways.

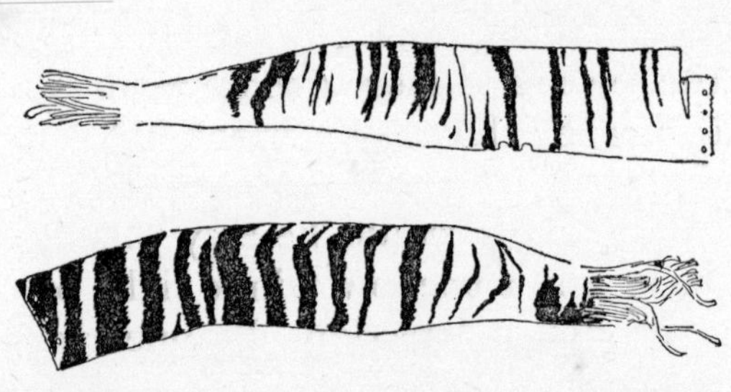

THE OKAPI—TWO BANDS OF HIDE

THE END

Sandy

A Gift For

2023

Tina

From

Published by Hallmark Gift Books,
a division of Hallmark Cards, Inc.,
Kansas City, MO 64141
Visit us on the Web at Hallmark.com.

Editorial Director: Theresa Trinder
Editor: Kara Goodier
Art Director: Chris Opheim
Designer: Laura Elsenraat
Production Designer: Dan Horton

ISBN: 978-1-63059-853-2
BOK1533

Made in China
0617

AMAZING AT EVERY AGE

BY SARAH MAGILL

YOU KNOW
THE OLD ADAGE IS TRUE:

IT'S NOT THE AGE, IT'S THE ATTITUDE.

Each year presents its own peculiar mix of problems and pleasures, excitement and unpredictability. Don't fight it. Embrace it. Then grab it by the hand and bring it into the next year, and the year after that, and keep on going.

Because you get to take it all with you. You get to revisit eighteen when you're eighty; you get to play at thirty when you're sixty-three. Maybe age is just a number, but it's *your* number. Play it for all it's worth.

So look forward. Look ahead. Reflect on what the years held. Dream about what they'll hold. It's a journey, becoming the woman you are. It takes years. You'll grow, you'll change, you'll be terrified, you'll be graceful, you'll be strong. You'll discover that every year you become a little more *you.*

And *you* are amazing.

18
IS IMMORTAL.

Eighteen knows it all and isn't afraid to tell you so. Everyone glows at eighteen. Eighteen is budding, even though you think you're in full bloom. Eighteen is charmingly self-involved, mercifully shortsighted, and fabulously in the moment. Eighteen decides to change the world, even if she doesn't know how long it takes to really do it. Dreams are enough to live on at eighteen—dreams and a reliable group of friends. Eighteen sees no conflict in proclaiming independence while letting your parents pay your car insurance.

Eighteen is coming alive.

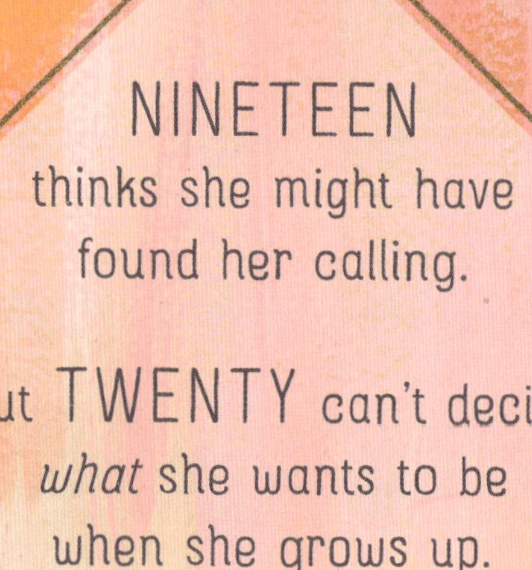

NINETEEN
thinks she might have
found her calling.

But TWENTY can't decide
what she wants to be
when she grows up.

21

IS WHEN OTHER PEOPLE START TO ACKNOWLEDGE HOW OLD AND WISE YOU ARE.

(The most important of these other people are bartenders.) Twenty-one is starting to get the real independence you thought you wanted. Twenty-one is learning to live with consequences, figuring out who you are (who knew you had so many faults!), and having a great time in spite of it all. Twenty-one is making decisions—real, adult decisions—and finding out there are no easy answers. Twenty-one is looking ahead, weighing the options, and choosing a path. Twenty-one is deciding that path wasn't quite right, backtracking, and taking another.

Twenty-one is a whole lotta hope.

TWENTY-TWO loves being carded.

TWENTY-THREE is wondering when
she's supposed to start feeling like an adult.

24

IS GETTING REALLY GOOD AT HER FIRST JOB.

25

IS LIVING THE

(WORKING-HARD, SLEEPING-LITTLE, SCRAPING-BY)

DREAM.

TWENTY-SIX

knows the difference
between a drinking buddy
and a lifelong friend.

TWENTY-SEVEN

is *really* falling in love
for the first time.

TWENTY-EIGHT loves a good happy-hour special.

TWENTY-NINE is learning to believe in herself.

30

IS WHEN YOU FINALLY KNOW SOME STUFF.

Not a lot, but at least enough to be able to pick out jeans that look good on you, friends who get you, and a job that doesn't drive you completely crazy. Thirty knows where she wants to go and has a plan (or is *planning* a plan) to get there. Thirty is realizing that you're not the center of the world and deciding whom you want in the center of yours. Thirty looks in the mirror and honestly likes at least part of what she sees. Thirty is missing the fun of your carefree, foolish pre-thirty nights but not the mornings after. At thirty, you know your limits. But you're just starting to test your strengths.

Thirty is becoming more you.

31

SOMETIMES JUST STAYS IN.

32
ASKS FOR
THAT RAISE.

THIRTY-THREE
plans to be in her prime
for the rest of her life.
THIRTY-FOUR
makes time for love, in all forms.

THIRTY-FIVE IS
LEARNING TO SAY NO
SO SHE'S GOT TIME
FOR THE REALLY
IMPORTANT STUFF.
35

THIRTY-SIX

IS STILL UP FOR A
RANDOM ADVENTURE (OR TWO).

THIRTY-SEVEN

NEEDS HER GIRLS.

THIRTY-EIGHT

IS REFINING HER TASTES,

BUT NOT TO THE POINT OF SNOBBERY.

THIRTY-NINE

SPEAKS UP.

40

IS GETTING THINGS DONE.

Forty makes good decisions—and after forty years of trial and error, Forty is pretty good at giving advice. Forty is seeing the stuff you wore in your twenties come back in style. Forty is being really good at what you do. . . and then questioning if it's what you really want to be doing anyway. Forty is putting your family (however you define it) first and asking them for help when you need it. Forty has earned the right to have some opinions. Forty is smart enough to pass for thirty and still feels twenty. Forty works hard and knows how to play hard, too. Forty celebrates goals and makes new ones.

Forty has stories.

FORTY-ONE can balance idealism and realism pretty well. **FORTY-TWO** doesn't waste her time trying to make other people happy anymore.

FORTY-THREE

has solidified her personal style but can't promise she won't change it again.

FORTY-FOUR

is still seeking, and that's OK.

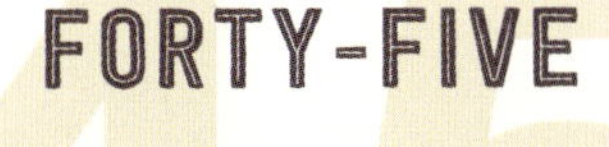

FORTY-FIVE

TAKES CARE OF BUSINESS, PEOPLE, AND HERSELF.

FORTY-SIX

is seeing a dream or two come true...

and in a way she never expected.

FORTY-SEVEN

finds time to give back.

FORTY-EIGHT practices mindfulness, even if it's still a struggle.

FORTY-NINE knows a shortcut.

50
WANTS EVERYBODY TO KNOW SHE'S FIFTY

because it makes how damn good she looks that much more impressive. Fifty lives what she believes. Fifty can argue you into the ground. Fifty has the dream job but is now reevaluating whether it's the *right* dream. Fifty has options. Lots of them. Fifty is the official get-back-in-shape age, otherwise known as the make-an-insane-fitness-goal-to-impress-your-friends age. Fifty gets the best birthday parties. Fifty always knows a guy. Fifty deserves a vacation. Fifty knows that the most important thing you can do in a day is tell the people you love that you love them.

Fifty has come a long way.

FIFTY-ONE

has her priorities straight—

and yes, fun is a priority.

FIFTY-TWO

is making new plans.

53

HAS BEEN THERE...
AND WANTS TO
GO AGAIN.

54
IS REALLY GOOD
AT WHAT SHE DOES.

55

CAN SMELL FREEDOM FROM THE NINE-TO-FIVE. . .

if she hasn't already bailed on a "traditional" job. Fifty-five is asking, "What's next?" Fifty-five is being amazed at everything you know and everything you still want to learn. Fifty-five has lifelong friends ready to take on the next decades together. Fifty-five won't apologize for anything because fifty-five always does the right thing. Fifty-five is turning that "hobby" into a second career. Fifty-five never waits to say the important stuff, like "Thank you" and "Yes, I'd like another glass of wine."

Fifty-five is really good at gratitude.

FIFTY-SIX

doesn't sweat the small stuff.

FIFTY-SEVEN

doesn't raise her hand—
she just says what
needs to be said.

FIFTY-EIGHT

is a really good mentor.

FIFTY-NINE

knows from experience
that experience is the best teacher.

"60"

DOESN'T CARE WHAT YOU THINK ANYMORE, AND THAT MAKES SIXTY DANGEROUS.

Sixty is confident, which makes Sixty undeniably sexy. Sixty means you can do whatever the hell you want (as long as you don't bother Seventy or Eighty). Sixty is eating well, exercising, and staying sharp. Sixty is making time to do all the things you said you didn't have time for before—reading those books, taking those trips, making those memories. Sixty is sharing the stuff you *thought* you knew at Eighteen but really know now. Sixty is living in the moment and making it count.

Sixty is happy.

SIXTY-ONE

NEGOTIATES LIKE A PRO.

SIXTY-TWO

KNOWS WHAT SHE WANTS *FOR REAL* NOW.

63

ISN'T AFRAID OF MAKING MISTAKES.

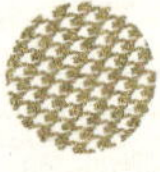

64

WORKS SMARTER—IF NOT HARDER—THAN EVER.

65

IS FORGETTING HOW OLD YOU ARE ON PURPOSE.

Sixty-five can tell when somebody's BS-ing. And when Sixty-five knows, they know she knows. Sixty-five knows retirement really means reframing, renewing, recharging, and recommitting. Sixty-five has enough practice to see the patterns in life, so Sixty-five usually knows what's coming. Sixty-five is the power of experience. Sixty-five is starting to make notes for that best-selling autobiography. Sixty-five is really good at this patience thing. Sixty-five has both weathered heartache and raced joy into next week. Sixty-five has some serious emotional endurance.

Sixty-five knows it and passes it on.

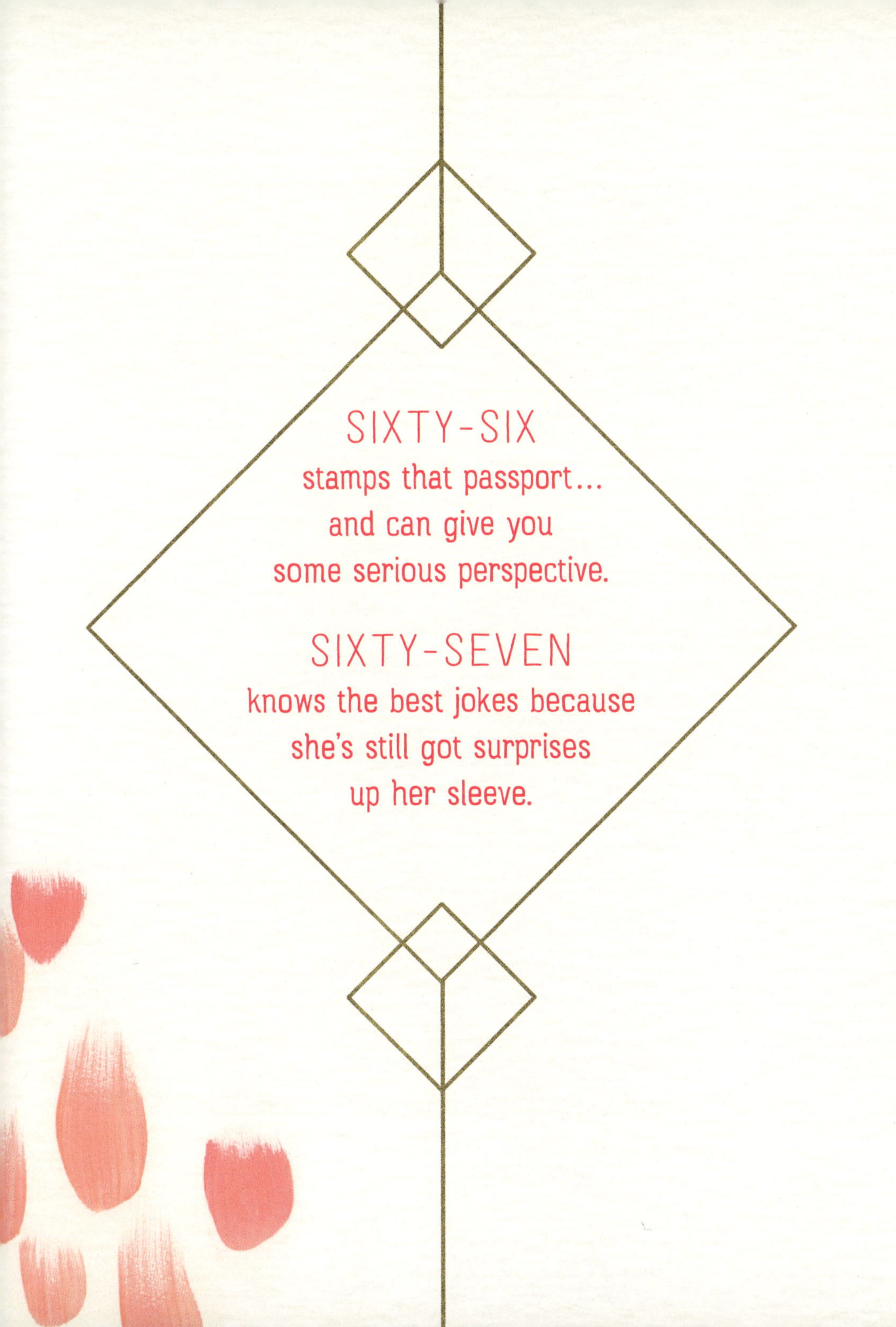

SIXTY-SIX

stamps that passport...
and can give you
some serious perspective.

SIXTY-SEVEN

knows the best jokes because
she's still got surprises
up her sleeve.

SIXTY-EIGHT sees right through you and loves you anyway.

SIXTY-NINE learns something new every day.

70

HAS PRIORITIES, AND SEVENTY DOESN'T MESS AROUND.

Seventy knows what Seventy wants to say—and has learned to say it well. Seventy is now officially a matriarch and rules with grace and a hearty sense of humor. Seventy knows life is both long and short, so Seventy forgives. Seventy looks ahead with clear eyes. Seventy takes the wheel. Seventy doesn't waste any more time trying to look younger. Instead, Seventy puts that precious energy toward her own particular brand of good. Seventy wields her strengths expertly and has learned to appreciate her weaknesses, too. Seventy is softer.

Seventy smiles more.

SEVENTY-ONE

gives her time and smarts
to the causes and people she believes in.

SEVENTY-TWO

is serious about having a good time.

SEVENTY-THREE

always finds time for the people she loves but has no room in her schedule for the boring stuff.

SEVENTY-FOUR

wears exactly what she wants.

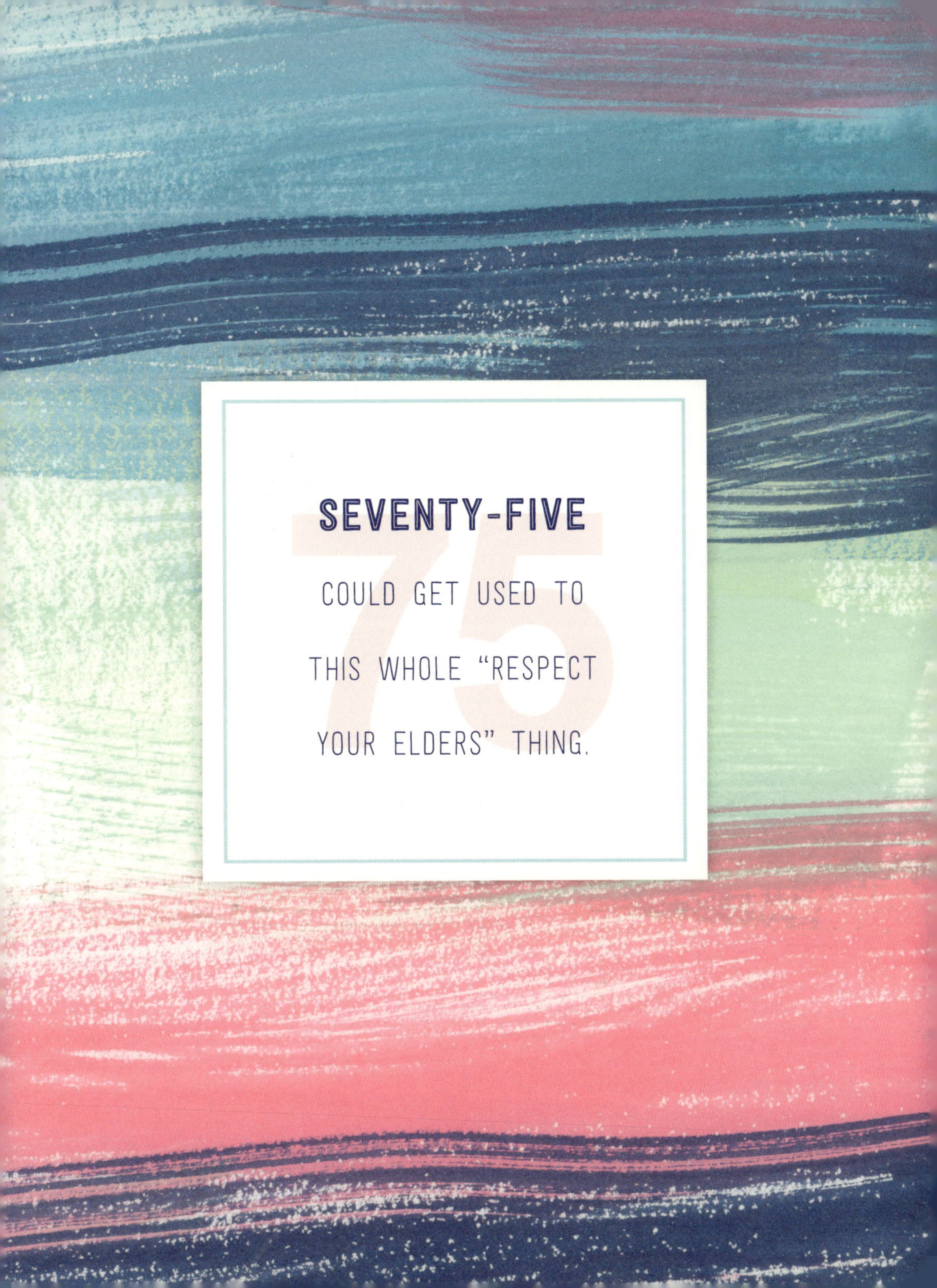
SEVENTY-FIVE
75
COULD GET USED TO
THIS WHOLE "RESPECT
YOUR ELDERS" THING.

76

EMBRACES THE MYSTERY.

77

WRITES HER OWN RULES.

SEVENTY-EIGHT

doesn't suffer fools,

excuses, or waiting rooms.

SEVENTY-NINE

proves glamour is ageless.

80

AMAZES.

Eighty starts passing things down and keeps paying it forward. Eighty thinks it's hilarious when people compliment her on how sharp she is. (They have no idea.) Eighty can see right through you, but Eighty has her secrets. Eighty doesn't take herself too seriously. Eighty doesn't take *anyone* too seriously. Eighty could go on and on, but Eighty will probably listen. Eighty has a family recipe that can't be replicated—no matter how hard you try. Eighty has had to fight like hell for *something*, and Eighty won. Eighty has a wide circle and always makes room for more.

Eighty finds goodness everywhere.

EIGHTY-ONE'S got a system, and it works for her.

EIGHTY-TWO doesn't believe in dieting.

EIGHTY-THREE
doesn't care so much about
who's right and who's wrong.
EIGHT-FOUR
rights regrets instead of
pretending she doesn't
have them.

EIGHTY-FIVE

RULES THE ROOST—

AND THE ROOST

IS WHEREVER SHE

HAPPENS TO BE.

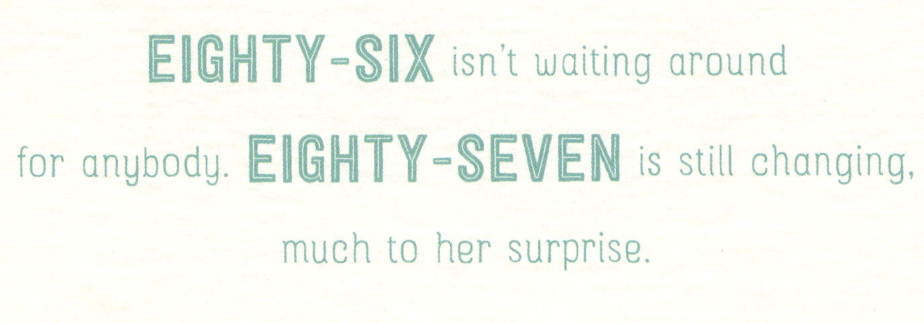

EIGHTY-SIX isn't waiting around for anybody. **EIGHTY-SEVEN** is still changing, much to her surprise.

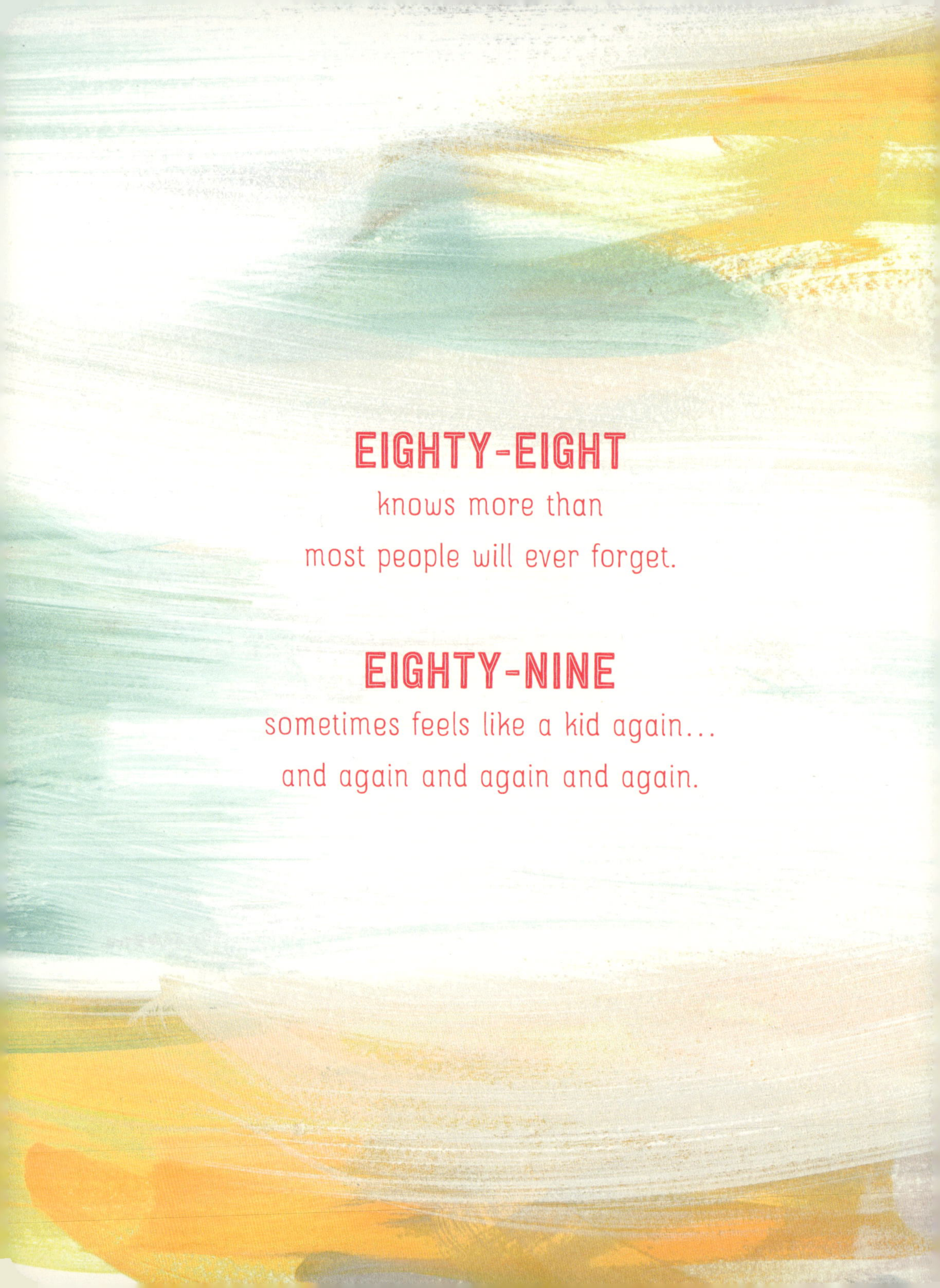

EIGHTY-EIGHT

knows more than
most people will ever forget.

EIGHTY-NINE

sometimes feels like a kid again...
and again and again and again.

90

HAS HAD YEARS TO PERFECT GRACEFUL SENIOR LIVING AND HAS IT DOWN TO A SCIENCE.

Ninety gets a lavish birthday bash every year from here on out. Ninety will tell you which cake she wants and exactly how to bake it (or where to buy it, for a discount). Ninety knows all the tricks. Ninety has decided delicacy is way overrated and just tells it like it is. Plus, Ninety knows you think it's funny to hear her hold forth on certain topics, and she's not above playing to a crowd. Ninety always goes first, and Ninety always gets seconds. If Ninety wants to preach, Ninety preaches. Ninety savors commanding respect.

Ninety earned it.

NINETY-ONE

loves surprise visitors.

NINETY-TWO

says, “Pull up a chair.”

NINETY-THREE lived it…and loved it.

NINETY-FOUR uses her stubbornness for good.

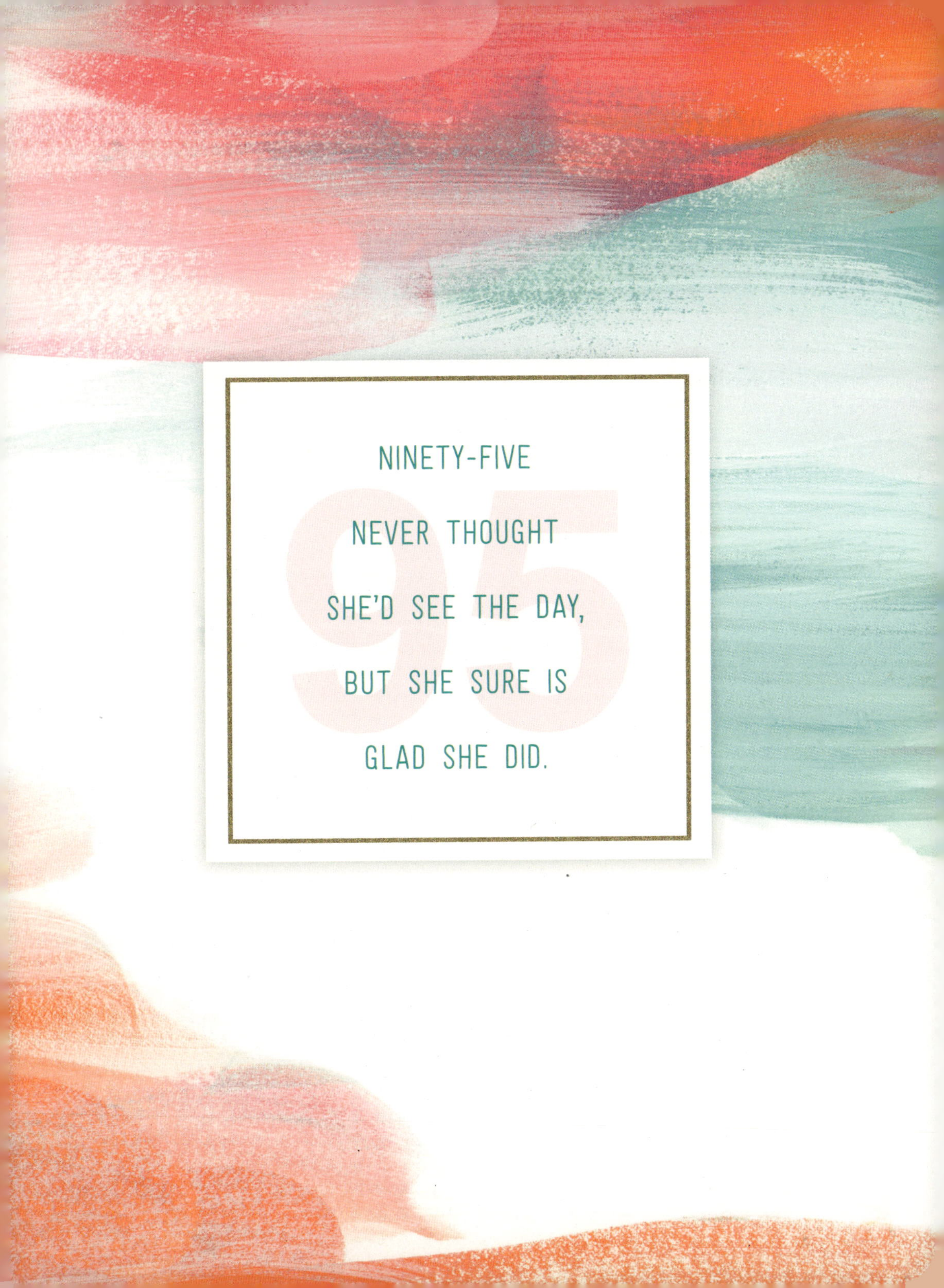
95
NINETY-FIVE
NEVER THOUGHT
SHE'D SEE THE DAY,
BUT SHE SURE IS
GLAD SHE DID.

96

GETS TO
THE POINT
AND SAVORS
EVERY SECOND.

97

HAS HEARD
IT BEFORE BUT
WOULD ENJOY
HEARING IT AGAIN.

NINETY-EIGHT
can't be shocked,
but she can be surprised.
NINETY-NINE
needs no introduction.

100
IS IMMORTAL.

One Hundred understands history because she helped write it. One Hundred has seen things. And One Hundred knows what they mean. One Hundred gets whatever One Hundred wants—and fast. One Hundred is lucky. One Hundred is thankful. One Hundred could have written this book. One Hundred may not dance a whole lot, but she still can groove. One Hundred parties every day. One Hundred gets interviewed. One Hundred accepts your awe, but One Hundred wants to hear your stories, too. One Hundred has most of the answers. But One Hundred still has questions—

questions and dreams.

IF YOU ENJOYED THIS BOOK OR IT HAS TOUCHED YOUR LIFE IN SOME WAY, WE'D LOVE TO HEAR FROM YOU.

Please write a review at Hallmark.com,

e-mail us at booknotes@hallmark.com,

or send your comments to:

Hallmark Book Feedback

P.O. Box 419034

Mail Drop 100

Kansas City, MO 64141